Environmental Engineering FE/EIT Preparation Sample Questions and Solutions

Copyright © Anthem Publishing®. All Rights Reserved

PO Box 1016

Tioga, ND 58852

Printed in the United States of America

March 2016

First Edition

Greetings!

I'd like to introduce you to Anthem Publishing. Anthem was born as an idea when a physics professor remarked to me, "Physics hasn't changed in 300 years, yet there is always enough change to update the textbook every semester." Anthem Publishing is built on the idea that high quality information should be affordable, because information is power.

Environmental Engineering FE/EIT Practice Problems and Solutions is a high quality book offered at a reasonable cost that will help you prep for the Fundamentals of Engineering Exam. Similar books retail for $80 to $120, while ours is available on Amazon for $25.

This book sets the standard for FE review boasting 110 practice problems with full solutions on topics such as Air Quality Engineering, Environmental Science and Management, Solid and Hazardous Waste Engineering, and Hydrologic and Hydrogeological Engineering.

At Anthem Publishing we believe that information is power, and high quality information should not only be available to those who can afford expensive textbooks. Our goal is to revolutionize the way information is presented in an academic setting, and help to educate the world.

Need more help studying for your exam? Check out www.FEexampleproblem.com, where we have compiled links to video reviews, study guides, and additional practice problems. Everything on our website is completely free! (Make sure and click on an ad or two!)

Cheers,

Anthem Publishing

Preface

This guide will be invaluable to help you prepare for the Fundamentals of Engineer Exam in the Environmental Engineering discipline. This guide was modeled in accordance with the exam specifications, and has problems representative of what you will encounter on the FE exam. Use this guide in conjunction with the NCEES FE Supplied Reference Handbook.

Errata

To report errata in this book, please email AnthemPublishing@yahoo.com.

1.) As a professional engineer originally licensed in 1982, you are asked to design a wastewater treatment plant for a growing, modern city. Ethically, you may only accept this offer if?

   a. You originally obtained a degree in Environmental Engineering.
   b. You have 20 years of experience in wastewater treatment systems.
   c. You are a member of a professional society of wastewater engineers.
   d. You are competent in the design of modern wastewater treatment plants.

2.) As a professional engineer, you are in charge of the remediation of a site contaminated by a historical release. A nearby landowner approaches you asking how the release and subsequent clean-up efforts will impact the health and wellbeing of his land. What, if anything, are you permitted to disclose to the landowner at this time?

   a. You may give your professional opinion when you have enough information to do so.
   b. Employer consent must be sought before giving any information to the concerned party.
   c. You may give the interested party any information that you deem applicable to the well-being of himself and his property.
   d. Being a licensed engineer requires you to disclose any and all information to the concerned party.

3.) As an employee of a company, you discover that a recently hired coworker does not meet all of the qualifications listed on his resume. This employee will not directly design or supervise any part of an upcoming project. What are you ethically responsible to do in this situation?

   a. You must inform your employer of this as it may impact current and future projects.
   b. As an employee you have no reason to involve yourself in in the processes of the company that you have no part in.
   c. Since the employee will not be designing or supervising any current projects, you have no ethical responsibilities.
   d. This matter should be discussed privately with the coworker, and a decision made at that point.

4.) If called to testify as an expert witness in a case determining liability for a contaminated site, a professional engineer should;

   a. Volunteer information on the background of the defendant.
   b. Offer opinion on the actions of the defendant.
   c. Provide to the best of his or her abilities, a technical analysis of his or her area of expertise.
   d. Offer speculation on why or how the defendant came to take certain actions.

5.) A recently finalized regulation impacts a design contract that you have already signed with a company the construction of a new copper smelter. In order to comply with the regulation, the cost on the project would need to be increased. What are you ethically obligated to do?

   a. Nothing, as the design for the facility is fluid.
   b. File an amendment to the bid detailing the impact of the regulation and subsequent cost increase after notifying the company.
   c. Add in the cost increase to the final invoice.
   d. Consider the smelter grandfathered in to previous regulations, and a redesign should not be necessary.

6.) Name the order of the differential equation seen below, and provide the general solution with the given boundary condition.

$$7y + \frac{dy}{dx} = 0 \qquad y(0) = 3$$

   a. 3rd Order, $y = \frac{3}{7} e^{-7t}$
   b. 1st Order, $y = -7e^{-3t}$
   c. 2nd Order, $y = e^{-7t}$
   d. 1st Order, $y = 3e^{-7t}$

7.) Find the value of the given interval.

$$\int_{\frac{\pi}{2}}^{\pi} 5 \cos z \, dz$$

a. 5
b. 0
c. -5
d. 10

8.) What is the magnitude of the cross product to the following vectors;
A= 3i+2j and B= -i+2j-k.

a. $\sqrt{24}$
b. 8
c. $\sqrt{77}$
d. 9

9.) What is the equation of the circle passing through points (x,y) of (-4,0), (0,0), and (0,4)?

a. $(x + 2)^2 + (y - 3)^2 = 18$
b. $(x + 2)^2 + (y - 2)^2 = 8$
c. $(x - 3)^2 + (y + 3)^2 = 18$
d. $(x + 3)^2 + (y + 9)^2 = 16$

10.) What is the value of $\ln(e^{6.7})$?

a. 2
b. 6.7
c. 1.99
d. 8.49

11.) From a group of 8 men and 4 women, a team of 4 will be formed by random selection. What is the probability that the team will consist of 4 men?

   a. 0.50
   b. 0.267
   c. 0.048
   d. 0.141

12.) The cycle times for an intermittent flare in a natural gas operation forms a normal distribution, with a standard deviation of 2 minutes. The mean cycle time is 11 minutes. The probability that a cycle selected at random will last more than 15 minutes is?

   a. 0.0179
   b. 0.9821
   c. 0.0228
   d. 0.0540

13.) What is the standard deviation of these four given values?

$$1,3,5,7$$

a. $\sqrt{5}$
b. 4
c. $\sqrt{6}$
d. 3.5

14.) What is the approximate probability that no two people in a randomly selected group of six having the same birthday?

a. 0.98
b. 0.94
c. 0.04
d. 0.96

15.) A haul truck has an initial cost of $100,000, and a salvage value of $20,000 after 10 years of use. Assuming straight line deprecation, what is the value of the haul truck after six years?

   a. $60,000
   b. $48,000
   c. $8,000
   d. $52,000

16.) A distiller can produce a bottle of gin for $5 that can be sold for $15. The factory has operating and maintenance costs of $80,000 per year and $100,000 per year of capital costs. How many bottles of gin must the company produce and sell to break even on the year?

   a. 18,000
   b. 21,000
   c. 300
   d. 45,000

17.) A crusher for use in a mining operation is expected to have a maintenance cost of $10,000 in the first year. The maintenance costs are expected to increase $1,000 per year. The interest rate is 8% per year, compounded annually. Over a five year period, what will be the approximate effective annual maintenance costs?

a. $12,500
b. $13,625
c. $11,850
d. $10,890

18.) A county is expecting the need for a new wastewater treatment system that will cost of $4 million in four years. If they were to invest funds today in an account earning 18% per year, how much would they need to invest?

a. $2,100,000
b. $6,100,000
c. $2,750,000
d. $3,200,000

19.) What is the value of 65 MPa in the following illustration called?

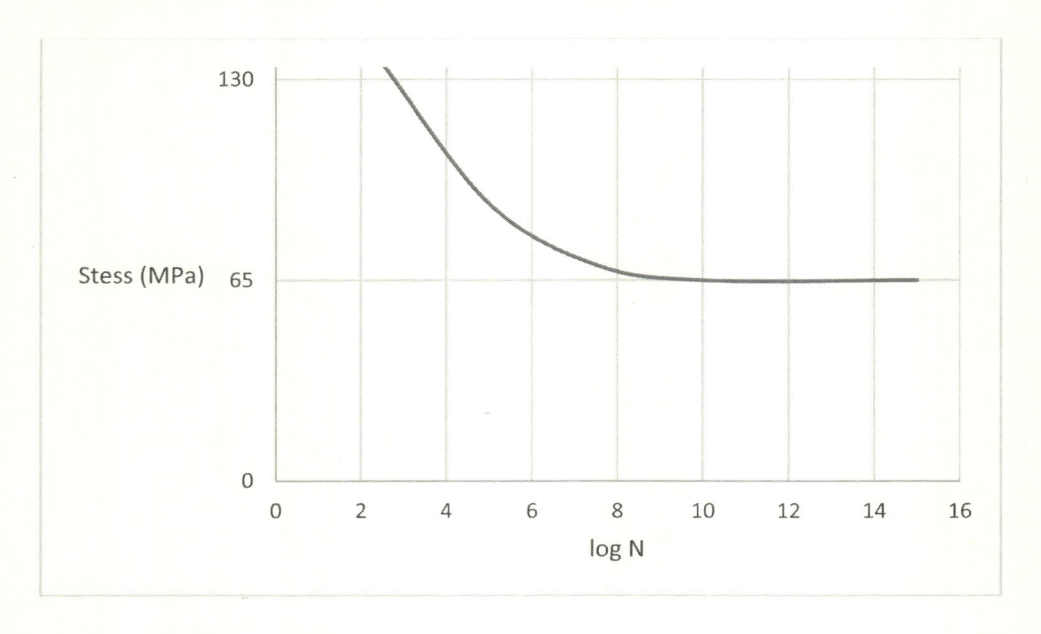

[Graph: Stress (MPa) vs log N, curve decreasing from 130 and leveling off at 65]

a. Fatigue Limit
b. Proportional Limit
c. Yield Point
d. Tensile Strength

20.) Galvanic action results from a difference in what property of metallic ions?

a. Half-cell potential
b. Specific heat
c. Defect Flux
d. Oxidation Potential

21.) While the corrosion of cast iron can be limited with a more electropositive coating, a coating that is less electropositive will lead to increased corrosion. Which of the following would be a poor choice to use as coating material on cast iron?

   a. Low carbon steel
   b. 2024 Aluminum Alloy
   c. Alcad 3S
   d. Silver

22.) A 1 Liter solution contains the following; 37 g H_2SO_4, 201 g $KMnO_4$, 13.8 g K_2SO_4, and 3.5 g Mn_2O_7. The equation for this reaction can be seen below. What is the equilibrium constant for this reaction?

$$H_2SO_4 + 2KMnO_4 \leftrightarrow K_2SO_4 + Mn_2O_7 + H_2O$$

   a. 0.621
   b. 0.002
   c. 1.0
   d. 0.006

23.) What volume of 2 M HCl is required to neutralize 20 mL of 5 M NaOH?

   a. 55 mL
   b. 37mL
   c. 50 mL
   d. 100 mL

24.) What is the empirical formula for a compound containing 62.16% Iron (Fe), 35.61% Oxygen (O), and 2.23% Hydrogen (H) by percent?

   a. $Fe(OH)_4$
   b. $Fe(OH)$
   c. $Fe_2(OH)$
   d. $Fe(OH)_2$

25.) The oxidation of chromium in a ground water supply is determined to follow first order kinetics, with a rate constant of 0.10 per day. If the resulting concentration is 5.6 mg/L after 10 days, what was the initial concentration? Assume the ground water acts as an ideal plug flow reactor.

   a. 32 mg/L
   b. 15 mg/L
   c. 65 mg/L
   d. 12 mg/L

26.) The following chemical equilibrium constant equation shows the stoichiometry of which of the following reactions?

$$K_{EQ} = \frac{[C]^3[D]^6}{[A]^1[B]^2}$$

   a. A+2B ↔ 3C+6D
   b. 2A+3B ↔ 3C+6D
   c. 3C+6D ↔ A+3B
   d. A+B₂ ↔ C3+D6

27.) Hydrogen Bromide (HBr) is neutralized with barium hydroxide by the equation seen below. If you have 25 mL of 0.1 M HBr, how much 0.05 M Ba(OH)₂ solution will be needed for neutralization by reaction shown below?

$$2HBr + Ba(OH)_2 \leftrightarrow BaBr_2 + 2H_2O$$

a. 30 mL
b. 20 mL
c. 25L
d. 25 mL

28.) What coefficients (a,b,c,d) balance the following equation?

$$aC_9H_{10}O + bO_2 \leftrightarrow cCO_2 + dH_2O$$

a. 1,14,9,10
b. 1,11,9,5
c. 2,23,19,10
d. 3,9,8,6

29.) Name the following compound.

$$CH_3 - CH_2 - CH_2 - \underset{\underset{CH_3}{|}}{\overset{\overset{CH_3}{|}}{C}} - CH = CH_2$$

a. 2,2-diethylpent-1-ene
b. 2-methylhex-2-ene
c. 3,3-diethylhex-2-ene
d. 3,3-dimethylhex-1-ene

30.) Name the following compound.

$$CH_3 - CH_2 - CH_2 - CH = CH - CH_3$$

a. Hex-2-ene
b. Hep-2-ene
c. Hex-4-ene
d. Pent-2-ene

31.) A general approach to organic chemistry reactions assumes that the characteristic properties of compounds are determined by what?

   a. Alkalinity
   b. Structure of the compound
   c. Alcoholic content
   d. Functional groups

32.) A stream has an initial BOD concentration of 500 mg/L and an initial dissolved oxygen deficit in the mixing zone of 2.2 mg/L. Assuming the reaeration of the water follows a rate constant of 0.6/day and the deoxygenation occurs at a rate of 0.7/ day, what is the dissolved oxygen concentration after 10 days according to the Steeter Phelps equation? The saturated dissolved oxygen concentration is 10 mg/L.

   a. 7.8 mg/L
   b. 4.5 mg/L
   c. 10 mg/L
   d. 5.5 mg/L

33.) How many grams of carbon dioxide is dissolved in 1 Liter of carbonated water if the manufacturer uses a pressure of 2 atm in the bottling process at 25°C? Note that the K_H of CO_2 in water is equal to 29.76 atm/(mol/L) at 25°C.

   a. 4.5 g
   b. 3.0 g
   c. 4.5 g
   d. 12 g

34.) A body of water that sees a large increase in the concentration of phosphates and nitrates from an influent waste stream will most likely be what?

   a. Eutrophic
   b. Oligotrophic
   c. Mesotrophic
   d. Nutrient Deficient

35.) What is the oxidation state of Chromium (Cr) in the following compound?

$$CrO_6^{-4}$$

a. 4
b. 7
c. 8
d. 6

36.) In a skin exposure toxicology study, the median single dose that is expected to kill 50 percent of a group of test animals is known as which of the following?

a. LC_{50}
b. Dosage
c. LD_{50}
d. Concentration

37.) A 70 kg man consumes 80 mg/day of soil that contains 20 mg/kg of benzene. He consumes this soil every day for 70 years. If the cancer slope factor for benzene is 0.029 kg*d/mg and the absorption factor is 1, his risk of developing cancer is most near which of the following?

a. 9.1×10^{-7}
b. 4.8×10^{-5}
c. 2.3×10^{-5}
d. 6.6×10^{-7}

38.) A 40 hour per week worker in an industrial facility is known to be exposed to a compound in the air at a concentration of 80 µg/m³. The worker inhales at a rate of 1 m³ per hour, and has a body mass of 70 kg. The chronic daily intake of this worker during an eight hour shift is which of the following? Assume the fraction of the chemical that is absorbed in the lungs is 0.65.

a. 0.006 mg/(kg*day)
b. 0.46 mg/(kg*day)
c. 0.04 mg/(kg*day)
d. 0.10 mg/(kg*day)

39.) During a study of dose and response, it is determined that the safe human dose of a chemical is 500 mg/day, and the no-observed-adverse-effect-level (NOAEL) for the chemical is 15 mg/kg*day. What uncertainty factor was used in the study, assuming it was conducted on average adult males?

   a. 2
   b. 18
   c. 4
   d. 6

40.) For carcinogens, the general acceptable value for risk in a population is less than which of the following?

   a. 10^{-2}
   b. 10^{-6}
   c. 10^{-7}
   d. 10^{-3}

41.) A man has been drinking water from the same rural well for 70 years. After analytical sampling takes placed it is determined that the water contains arsenic at concentrations of 5 mg/L. If the cancer slope factor of arsenic is 0.67 kg*day/mg, what is his cancer risk from this exposure? Assume an ingestion rate of 2L per day.

   a. 0.15
   b. 0.10
   c. 0.02
   d. 1 X 10^{-6}

42.) Which of the following signifies the dosage below which there are no harmful effects?

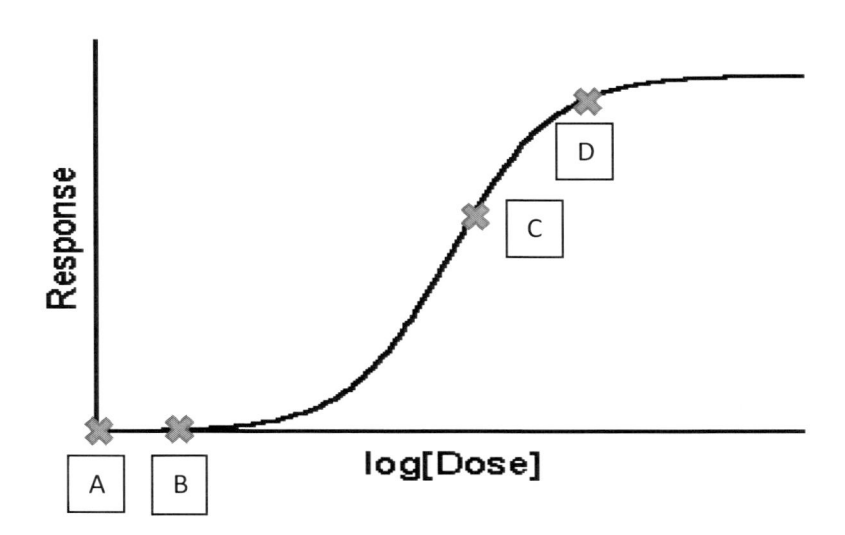

   a. A
   b. B
   c. C
   d. D

43.) A fluid with a vapor pressure of 0.4 Pa and a specific gravity of 10 is used in a barometer. If the fluids column height is 1 m, what is the atmospheric pressure?

   a. 98 KPa
   b. 114 Kpa
   c. 76 KPa
   d. 84 KPa

44.) A pressure of 1000 lbf/ft² is measured 10 feet below the surface of an unknown liquid. What is the specific gravity of the liquid?

   a. 1.60
   b. 1.25
   c. 30.3
   d. 0.35

45.) A six inch plastic pipe that is 1200 feet long has a Hazen Williams coefficient of 120. If the pipe is carrying 0.6 cfs of water and the pipe is full, the head loss in feet of water is most nearly?

a. 0.8
b. 2.5
c. 9
d. 15

46.) A sanitary sewer system with a depth to diameter ratio of 0.6 has a flow capacity of 50 cfs. Assuming the roughness coefficient varies with depth, what is the designed flow rate for this system?

a. 12 cfs
b. 50 cfs
c. 46 cfs
d. 28 cfs

47.) A sanitary sewer system is being designed for an apartment complex. The calculations show a flow rate of 1.57 m³/s, a Manning's roughness coefficient of 0.014 and a diameter of 1 meter. What slope percent was used in the design?

   a. 0.25%
   b. 2.5%
   c. 0.05%
   d. 0.5%

48.) The critical depth of a rectangular stream with a cross sectional area of 7 m² and a width of 3 m is most nearly?

   a. 2.33 m
   b. 1.76 m
   c. 1.54 m
   d. 2.48 m

49.) Five cubic feet per second of water is being pumped through 1000 feet of eight inch pipe. Neglecting minor losses, the pump will need to be able to handle how many feet of total head? The static head is 20 feet, and the friction coefficient is 140.

    a. 91 feet
    b. 83 feet
    c. 74 feet
    d. 45 feet

50.) A pump station includes two centrifugal pumps in parallel. The pump curve for a single pump can been seen below.

$$Head = 70 - 0.002Q^2$$

Where Q is in units of gpm, and the equation is valid from $50\ gpm < Q < 300\ gpm$

The piping system needs to overcome a static head of 40 feet. The friction loss for the system is defined by the following equation;

$$Friction\ Loss\ (feet) = 0.0004Q^2$$

Where Q is in units of gpm, and the equation is valid from $50\ gpm < Q < 300\ gpm$

When the system is operating a single pump, the flow rate is closest to which value?

    a. 122 gpm
    b. 150 gpm
    c. 112 gpm
    d. 130 gpm

51.) A rectangular suppressed weir is being sized to handle a flow rate of 3 cubic meters per second. If the depth of the water in the weir needs to be less than 1 meter, how long does the weir need to be?

   a. 1.8 m
   b. 1.6 m
   c. 3.2 m
   d. 1 m

52.) The orifice shown below is discharging freely into the atmosphere.

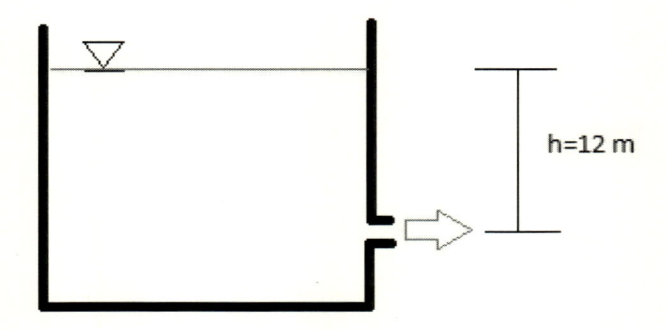

h=12 m

If the coefficient of discharge is 0.34 and the diameter of the discharge pipe is 0.25m, what is the resulting flow rate?

   a. 0.20 m³/sec
   b. 0.15 m³/sec
   c. 0.12 m³/sec
   d. 0.26 m³/sec

53.) A hose shoots out a jet of water vertically with a velocity (v) and a flow rate (Q). A horizontal plate is located is located directly above the nozzle at a height (h). The density of the water is ρ. What is the force necessary to keep the plate in equilibrium against the force of the water jet?

- a. $Q\rho\sqrt{v^2 - 2gh}$
- b. $\sqrt{v^2 2gh}$
- c. $Q\rho\sqrt{v - 2gh}$
- d. $\frac{1}{2}mv_f^2$

54.) An oil pipeline pumps 1000 barrels of crude oil per minute with a specific gravity of 0.68. The static head on the system is 40 feet. Neglecting minor losses, how much power must be applied by a pump assuming 100% efficiency?

- a. 425 hp
- b. 290 hp
- c. 300 hp
- d. 255 hp

55.) Air is compressed in a system to 1/5 of its initial volume. The final temperature is 500°C, and the process is frictionless and adiabatic. What is the initial temperature?

   a. 138°C
   b. 124°C
   c. 102°C
   d. 94°C

56.) A refrigeration system is said to operate in a Carnot cycle. The system receives heat from a reservoir at 0°C. If the coefficient of performance "COP" for the system is 3, what is the power input per ton of refrigeration?

   a. 1355 W/ton
   b. 11,500 W/ton
   c. 1285 W/ton
   d. 1180 W/ton

57.) An air sample taken has a temperature of 25°C and a relative humidity of 30%. What temperature is the dew point closest to?

   a. 10°C
   b. 15°C
   c. 5°C
   d. 0.01°C

58.) What are the products of complete combustion of gaseous hydrocarbons?

   a. Only Carbon Monoxide
   b. Only Carbon Dioxide
   c. Carbon Dioxide, Carbon Monoxide, and Water
   d. Carbon Dioxide and Water

59.) A city with a population of 40,000 people has an average water usage of 180 gallons/person per day. If the return rate is 75%, what is the maximum daily flow rate for wastewater?

 a. 5.4 MGD
 b. 14.6 MGD
 c. 2.2 MGD
 d. 11 MGD

60.) A small town is growing at an exponential rate of 2.2% per year. If the per capita water usage is 150 gallons per day, and the town is planning a drinking water plant to last for 20 years, what capacity should the plant be designed for? The current population is 5,500.

 a. 1.3 MGD
 b. 0.8 MGD
 c. 0.5 MGD
 d. 1.8 MGD

61.) A metropolitan area is growing at a rate of 3% per year. If this is assumed to be a linear growth rate, what will be the population in 3 years if the current population is 1.2 million?

   a. 1.4 million
   b. 1.7 million
   c. 1.3 million
   d. 1.9 million

62.) A storm produced 2 inches of water in 30 minutes. What is the probability of a storm of this intensity occurring during a given year according to the following graph?

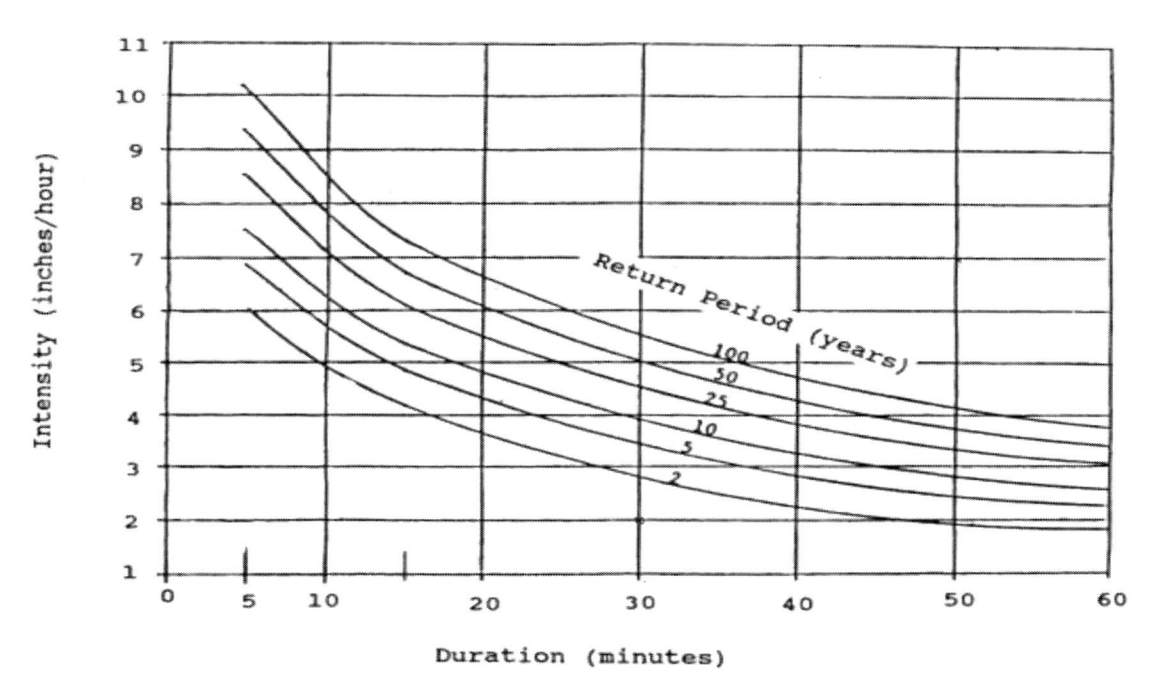

   a. 0.10
   b. 0.50
   c. 0.02
   d. 0.01

63.) A watershed with an area of 160 acres used to be pasture land with a runoff coefficient of 0.15. The area was developed into a subdivision with a runoff coefficient of 0.35. During similar rain events, an increase in runoff of 96 cfs was measured in the developed area. What was the intensity of the rainfall event?

   a. 2 in/hr
   b. 3 in/hr
   c. 4 in/hr
   d. 5 in/hr

64.) If a paved concrete parking lot is designed, storm water best management practices could be used to do which of the following?

   a. Increase infiltration
   b. Increase time of concentration
   c. Increase peak discharge
   d. Decrease rainfall intensity

65.) What is the runoff in acre-feet for a 640 acre basin with a maximum basin retention of 0.5 inches for a 1 inch storm event?

   a. 35 acre-feet
   b. 30 acre-feet
   c. 25 acre-feet
   d. 20 acre-feet

66.) A storm water conveyance channel should be used to deal with which type of flow?

   a. Concentrated flow
   b. Sheet flow
   c. Nutrient rich flow
   d. Base flow

67.) The chart below shows the daily usage of water from a reservoir. What minimum volume should the reservoir be designed to store?

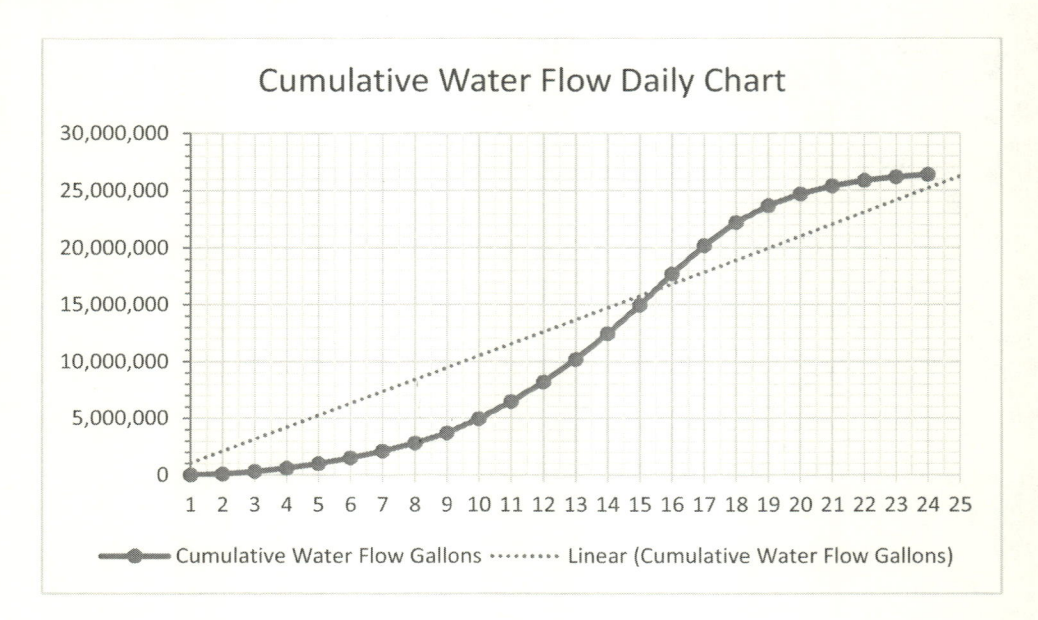

Cumulative Water Flow Daily Chart

— Cumulative Water Flow Gallons ········· Linear (Cumulative Water Flow Gallons)

a. 12 MG
b. 26 MG
c. 18 MG
d. 9 MG

68.) Excessive hardness of water in a potable water system can cause many issues. Potable water can be treated with which of the following to reduce hardness?

a. $Ca(OH)_2$
b. HCl
c. H_2SO_4
d. $(HCO_3)_2$

69.) A raw water source flows into a circular tank where it is treated. The required hydraulic residence time is 1 hour, and the flow rate in is 100 gallons per minute. If the tank has a maximum height of 10 feet, what is the required diameter?

   a. 8 feet
   b. 5 feet
   c. 10 feet
   d. 12 feet

70.) A typical primary clarifier is needed to remove 64% of the suspended solids of the treated water. If the flow rate is 1 cfs, what is the necessary surface area of the clarifier?

   a. 940 ft^2
   b. 810 ft^2
   c. 1100 ft^2
   d. 850 ft^2

71.) A filtration system is used to treat water. The water has a terminal settling velocity of 13 m/s, and the porosity of the fluidized bed is 0.1. At what velocity should the system be backwashed?

   a. 0.10 mm/s
   b. 1.00 mm/s
   c. 0.40 mm/s
   d. 0.60 mm/s

72.) A turbulent flow impeller is used in a rapid mix water treatment system. If the turbine has 6 curved blades and operates at a rotational speed of 100 rpm, how much power is needed to run the turbine? The diameter of the impeller is 0.2 m, the water density is 1000 kg/m³, and the flow is turbulent.

   a. 7 W
   b. 8 W
   c. 10 W
   d. 68 W

73.) A potable water disinfection treatment is described by first order kinetics with a rate constant of 0.15 per minute. An ideal CMFR reactor is used, with a hydraulic residence time of 30 minutes. What is the effluent concentration if the initial concentration is 1000 mg/L?

   a. 180 mg/L
   b. 1000 mg/L
   c. 225 mg/L
   d. 150 mg/L

74.) A conventional activated sludge system treats 6 MGD. The aeration channel has a cross section of 30 ft by 11 ft. If the hydraulic residence time is 5 hours, what is the required length of the channel?

   a. 600 ft
   b. 700 ft
   c. 650 ft
   d. 500 ft

75.) An activated sludge aeration tank is 100 feet long, and 20 feet deep. The influent flow rate is 1.5 MGD, and the influent BOD is 160 mg/L. If the volumetric BOD_5 loading rate is 100 lb/(1000 ft^2*day), what is the necessary width of the tank?

   a. 350 feet
   b. 200 feet
   c. 500 feet
   d. 180 feet

76.) What is the solids loading rate in lb/day*ft2 for activated sludge clarifiers with the following characteristics?

Number of units in parallel=2
Unit Diameter= 120 ft
Unit side water depth= 15 ft
Raw waste water inflow rate= 10 MGD
Return Activated Sludge flow rate= 3.0 MGD/clarifier
MLSS= 3500 mg/L
Raw Water BOD_5 = 225 mg/L
Raw Water SS = 265 mg/L

   a. 20
   b. 18
   c. 16
   d. 7

77.) A waste water treatment lagoon is designed to treat 1 MGD of wastewater with a concentration of 200 mg/L BOD$_5$ down to a concentration of 10 mg/L. The lagoon can be considered a second order ideal CMFR, with a rate coefficient of 3/day. What is the necessary volume of the lagoon?

   a. 4,350,000 ft^3
   b. 107,000 ft^3
   c. 85,000 ft^3
   d. 600,000 ft^3

78.) An experimental air stripping water treatment system is designed with a liquid mole loading rate of 350 Kmol/s*m^2, and a height of 10 meters. What is the overall transfer rate coefficient?

   a. 1.12/sec
   b. 0.68/sec
   c. 0.87/sec
   d. 0.63/sec

79.) Determine the length of a rectangular clarifier if the flow rate is 5 MGD, the retention time is 2 hours, and the tank has a depth of 10 feet. The Length: Width ratio is 3:1.

   a. 112 feet
   b. 130 feet
   c. 108 feet
   d. 100 feet

80.) Determine the depth of the sorption zone on an activated carbon system with a total carbon depth of 10 feet, a volume of 2 MG at breakthrough, and a volume of 2.5 MG at exhaustion.

   a. 2.5 feet
   b. 2.2 feet
   c. 3 feet
   d. 3.3 feet

81.) Determine the minimum acreage needed to handle an inflow loading of 2750 lb BOD/day given the following design constraints; BOD max loading rate of 35 lb BOD/day/acre, and 0.2 lb BOD per 1000 ft³ with a max depth of 6 feet.

   a. 70 acres
   b. 34 acres
   c. 53 acres
   d. 79 acres

82.) An air sample shows a H₂S value of 75 ppm. The corresponding concentration in µg/L at 25°C and 1 atm of pressure is which of the following?

   a. 104
   b. 78
   c. 121
   d. 10

83.) Secondary air quality pollutants are primarily formed from the reactions of primary air quality pollutants. Where do these reactions generally occur?

   a. Combustion processes
   b. Atmosphere
   c. Pollution Control devices
   d. Production processes

84.) Coal that enters a power plant is pretreated to remove 98% of the mercury. If the coal has an average concentration of 1 g Hg/ton and the input rate for the power plant is 40,000 tons/hour, how much mercury is being emitted in g/s?

   a. 0.11 g/s
   b. 0.38 g/s
   c. 800 g/s
   d. 0.22 g/s

85.) An incinerator has been tested and confirmed to remove 98% of all VOC's. The outlet concentration must be less than 50 µg/m³ and the maximum volumetric flow rate for the system is 10 m³/sec. What is the maximum flow rate of VOC's into the system in kg/day?

   a. 1.3
   b. 2.2
   c. 5.3
   d. 6.8

86.) The stability class associated with the healthiest ambient air quality events is which of the following?

   a. Stable
   b. Slightly stable
   c. Slightly unstable
   d. Very unstable

87.) A stack from an ore roasting operation emits CO at a rate of 20 g/s. The effective stack height is 100 m. What is the approximate maximum ground level concentration at a distance of 2 km downwind if the wind is blowing at 2.5 m/s on a sunny day with a few broken clouds?

    a. 355 µg/m³
    b. 2.5 µg/m³
    c. 61 µg/m³
    d. 34 µg/m³

88.) Which pollutant would tend to travel the furthest in a plume?

    a. PM 2.5
    b. PM 10
    c. PM 25
    d. PM 100

89.) A 40 foot tall scrubber treats a gas stream to a concentration of 100 µg/m³. If the height of a transfer unit is 2 feet and it has a stripping factor of 0.98, what is the inlet concentration of the gas stream?

   a. 2000 µg/m³
   b. 2250 µg/m³
   c. 1750 µg/m³
   d. 20,000 µg/m³

90.) A high efficiency air cyclone has a body diameter of 0.5 m. It treats an air flow of 120 m³/min, at 350 K and 1 atm. What is the number of effective turns of the cyclone?

   a. 4
   b. 5
   c. 6
   d. 07

91.) An Electrostatic Precipitator (ESP) at a coal fired power plant must stay above 97% efficiency to stay in compliance. The ESP has 3 plates measuring 5 m by 10 m each, and the terminal drift velocity of the particles is 10 cm/s. What is the maximum allowable actual gas flow rate?

   a. 2.6 m³/s
   b. 8.6 m³/s
   c. 14 m³/s
   d. 25.4 m³/s

92.) Air flow in a baghouse at a cement plant has a temperature of 25°C and a flow rate of 10,000 scfm. Woven fabric bags are used to remove particulates from the air. Each bag is 9 inches in diameter, and 10 feet long. How many bags are required to filter the air?

   a. 216 bags
   b. 188 bags
   c. 143 bags
   d. 233 bags

93.) A wood treatment plant operates 360 days per year. The permit specifies that no more than 25 tons per year may be emitted of VOC's. If the emission factor is 5.8 X 10^{-3} lb/ft^3 of treated wood, how many cubic feet can be treated in one year assuming no control devices are present?

   a. 6.4 million
   b. 7.8 million
   c. 9.2 million
   d. 8.6 million

94.) If an environmental assessment (EA) is performed for a proposed project and it is determined to have the possibility to cause significant impact, what is the next step in the permitting process?

   a. Environmental Impact Study
   b. Remediation
   c. Delineation
   d. Baseline sampling

95.) A community generates 75,000 lb/day of solid waste that is deposited in a municipal landfill. The permit for the landfill stipulates a ratio of refuse to cover of 1:2. The in place density of the material (both refuse and cover) is 1200 lb/yd³. If the operational permit is for 10 years, how much material (yd³) does the landfill need to be designed to hold?

a. 684,000
b. 732,000
c. 61,000
d. 183,000

96.) A landfill is limited to a depth of 60 feet. The landfill serves a community of 110,000 people, with a waste generation rate of 4 lb/day per person. If the density of the compacted waste is 1000 lb/yd³ and the cover material account for 20% of the total volume, what amount of acreage must be disturbed? Assume a 20 year landfill life.

a. 3.3 acre
b. 49 acre
c. 42 acre
d. 6.5 acre

97.) Which of the following class of compounds could be safely combined with hydrazine?

   a. Carbamates
   b. Esters
   c. Caustics
   d. Cyanides

98.) An industrial facility has the following hazardous wastes on site; water reactive substances, aromatic hydrocarbons, amides, and cyanides. What is the minimum number of containers needed to store the waste?

   a. 1
   b. 2
   c. 3
   d. 4

99.) A clay liner is being installed in a landfill and will be compacted to a porosity of 0.20, a hydraulic conductivity of 10^{-5} ft per day, and a thickness of 5 feet. If the hydraulic head is 15 feet, what is the breakthrough time in years?

   a. 88
   b. 94
   c. 69
   d. 54

100.) Which of the following is a characteristic that is used to classify waste as hazardous?

   a. Oxidative
   b. Inertivity
   c. Reactivity
   d. Leachability

101.) After initial compaction waste at a landfill has a density of 800 lb/yd³. After the waste is placed in the landfill, cover material is placed on top until the overburden pressure on the waste is 200 psi. What is the specific weight of the material at this pressure? Assume empirical constants of 0.1 yd³/in², and 5 X 10⁻⁴ yd³/lb.

   a. 1800 lb/yd³
   b. 1600 lb/yd³
   c. 1900 lb/yd³
   d. 1300 lb/yd³

102.) A cover layer on a landfill has undergone a change in the amount of water that it holds in storage. The change is measured by saturation testing as a decrease of 2.0 inches per unit area. If the layer has undergone precipitation of 3 inches, runoff of 1.2 inches, and evapotranspiration of 0.3 inches, how much water per unit area has percolated into the waste below?

   a. 1.5 in
   b. 2.0 in
   c. 2.5 in
   d. 3.5 in

103.) The activity of a radionuclide was measured after 30 days at 600 Bq. If the half-life is known to be 11 days, what was the initial activity level?

a. 100 Bq
b. 4000 Bq
c. 2000 Bq
d. 3000 Bq

104.) A saturated clay loam soil has a porosity of 35%. If the specific gravity of the clay is 2.80, what is the specific weight of the saturated soil?

a. 163 lb/ft^3
b. 135 lb/ft^3
c. 120 lb/ft^3
d. 118 lb/ft^3

105.) Using the Dupuit formula, what flow rate from a well (cfs) is required to draw down an aquifer 20 feet from the base of the aquifer a quarter mile away? It is determined that the well draws down the aquifer 8 feet from the base an 1/8th of a mile away at the same flow rate. Assume K is equal to 10^{-3} ft/sec.

    a. 1.5 cfs
    b. 18 cfs
    c. 12 cfs
    d. 0.09 cfs

106.) If a partially confined aquifer has a hydraulic conductivity of 3×10^{-2} cm/s, a thickness of 35 m, and a storage coefficient of 0.20, what is the transmissivity of the aquifer?

    a. 585 m²/day
    b. 1115 m²/day
    c. 680 m²/day
    d. 910 m²/day

107.) A confined aquifer is 30 m thick and has a width of 0.5 km. Two observation wells are located 0.8 km apart in the direction of flow. The head in well #1 is 50 m and in well #2 it is 38 m. Hydraulic conductivity is 0.5 m/day in the aquifer and the effective porosity is 0.3. What is the flow rate of water in the aquifer?

a. 115 m³/day
b. 95 m³/day
c. 85 m³/day
d. 105 m³/day

108.) An aquifer drops 2 feet in elevation over a distance of 1350 feet. A tracer is deployed in the aquifer, and it takes 8 months to travel 300 feet. What is the hydraulic conductivity of the aquifer in ft per second?

a. 10^{-1}
b. 10^{-2}
c. 10^{-3}
d. 10^{-4}

109.)   An aquifer has a thickness of 50 feet, over an area of 10 acres. If the porosity of the aquifer is 0.4 and the saturation is 90%, how much water is contained in the aquifer?

   a. 170 acre-ft
   b. 180 acre-ft
   c. 190 acre-ft
   d. 225 acre-ft

110.)   In which part of the groundwater system are the pore spaces not filled with water?

   a. Capillary Fringe
   b. Zone of Aeration
   c. Zone of saturation
   d. Hydraulic gradient

Answer Key

1.	D	31.	D	61.	C	91.	B
2.	B	32.	B	62.	A	92.	A
3.	A	33.	B	63.	B	93.	D
4.	C	34.	A	64.	B	94.	A
5.	B	35.	C	65.	B	95.	A
6.	D	36.	C	66.	A	96.	C
7.	C	37.	D	67.	D	97.	C
8.	C	38.	A	68.	A	98.	B
9.	B	39.	A	69.	C	99.	C
10.	B	40.	B	70.	B	100.	C
11.	D	41.	B	71.	C	101.	A
12.	C	42.	B	72.	A	102.	D
13.	A	43.	A	73.	A	103.	B
14.	D	44.	A	74.	D	104.	B
15.	D	45.	C	75.	B	105.	A
16.	A	46.	D	76.	A	106.	D
17.	C	47.	D	77.	C	107.	A
18.	A	48.	A	78.	D	108.	C
19.	A	49.	A	79.	B	109.	B
20.	D	50.	C	80.	B	110.	B
21.	D	51.	B	81.	D		
22.	B	52.	D	82.	A		
23.	C	53.	A	83.	B		
24.	C	54.	B	84.	D		
25.	B	55.	A	85.	B		
26.	A	56.	D	86.	D		
27.	D	57.	C	87.	D		
28.	B	58.	D	88.	A		
29.	D	59.	B	89.	C		
30.	A	60.	A	90.	C		

1.) As a professional engineer originally licensed in 1982, you are asked to design a wastewater treatment plant for a growing, modern city. Ethically, you may only accept this offer if?

   a. You originally obtained a degree in Environmental Engineering.
   b. You have 20 years of experience in wastewater treatment systems.
   c. You are a member of a professional society of wastewater engineers.
   d. **You are competent in the design of modern wastewater treatment plants.**

   Correct Answer (D)

   Solution: Referencing the Fundamentals of Engineering Supplied-Reference Handbook; Ethics section, under the Licensee's Obligations to Employer and Clients it states "Licensees shall undertake assignments only when qualified by education or experience in the specific technical fields of engineering or surveying involved." Therefore, the correct answer is D, an engineer must be competent in the area in which he or she is designing.

2.) As a professional engineer, you are in charge of the remediation of a site contaminated by a historical release. A nearby landowner approaches you asking how the release and subsequent clean-up efforts will impact the health and wellbeing of his land. What, if anything, are you permitted to disclose to the landowner at this time?

   a. You may give your professional opinion when you have enough information to do so.
   b. **Employer consent must be sought before giving any information to the concerned party.**
   c. You may give the interested party any information that you deem applicable to the well-being of himself and his property.
   d. Being a licensed engineer requires you to disclose any and all information to the concerned party.

   Correct Answer (B)

   Solution: Referencing the Fundamentals of Engineering Supplied-Reference Handbook; Ethics section, under Licensee's Obligation to Employer and Clients it states "Licensees shall not reveal facts, data, or other information obtained in a professional capacity without the prior consent of the client or employer except as authorized to by state law." In this situation you must seek employer consent before releasing information to the concerned landowner.

3.) As an employee of a company, you discover that a recently hired coworker does not meet all of the qualifications listed on his resume. This employee will not directly design or supervise any part of an upcoming project. What are you ethically responsible to do in this situation?

   a. **You must inform your employer of this as it may impact current and future projects.**
   b. As an employee you have no reason to involve yourself in in the processes of the company that you have no part in.
   c. Since the employee will not be designing or supervising any current projects, you have no ethical responsibilities.
   d. This matter should be discussed privately with the coworker, and a decision made at that point.

   Correct Answer (A)

   Solution: Referencing the Fundamentals of Engineering Supplied-Reference Handbook; Ethics section, Licensee's Obligation to other Licensees it states "Licensees shall not falsify or permit misrepresentation of their, or their associates, academic or professional qualifications. They shall not misrepresent or exaggerate their degree or responsibility in prior assignments nor the complexity of said assignments." It is your ethical obligation to inform your employer that the employee had violated ethical standards, even though he or she is not directing design or supervise parts or the upcoming project. The employee may be put into a design or supervisory role on the next project and make a crucial mistake.

4.) If called to testify as an expert witness in a case determining liability for a contaminated site, a professional engineer should;

   a. Volunteer information on the background of the defendant.
   b. Offer opinion on the actions of the defendant.
   c. **Provide to the best of his or her abilities, a technical analysis of his or her area of expertise.**
   d. Offer speculation on how or why the defendant came to take certain actions.

   Correct Answer (C)

   Solution: Referencing the Fundamentals of Engineering Supplied-Reference Handbook; Ethics section it states, "Licensees shall be objective and truthful in professional reports, statements, or testimony." An engineer must not speculate, offer opinion, or volunteer information on the background of a defendant.

5.) A recently finalized regulation impacts a design contract that you have already signed with a company the construction a new copper smelter. In order to comply with the regulation, the cost on the project would need to be increased. What are you ethically obligated to do?

   a. Nothing, as the design for the facility is fluid.
   b. **File an amendment to the bid detailing the impact of the regulation and subsequent cost increase after notifying the company.**
   c. Add in the cost increase to the final invoice.
   d. Consider the smelter grandfathered in to previous regulations, and a redesign should not be necessary.

   Correct Answer (B)

   Solution: Referencing the Fundamentals of Engineering Supplied-Reference Handbook; Ethics section it states, "Licensees, in the performance of their services for clients, employers, and customers, shall be cognizant that their first and foremost responsibility is to the public welfare." The design engineer must inform the company of the impact of the regulation, and file an amendment to the bid to ensure public welfare is protected. This is also a good business practice to adhere to.

6.) Name the order of the differential equation seen below, and provide the general solution with the given boundary condition.

$$7y + \frac{dy}{dx} = 0 \qquad y(0) = 3$$

   a. 3rd Order, $y = \frac{3}{7} e^{-7t}$
   b. 1st Order, $y = -7e^{-3t}$
   c. 2nd Order, $y = e^{-7t}$
   d. **1st Order, $y = 3e^{-7t}$**

   Correct Answer (D)

   Solution: Referencing the Fundamentals of Engineering Supplied-Reference Handbook; Mathematics Section; Differential Equations.

   The equation given is a First Order Linear Homogenous Differential Equation with constant coefficients. Therefore;

$$y' + ay = 0 \; where \; a = real \; constant$$

$$\text{Solution: } y = Ce^{-at} \text{ where}$$

$$C = \text{Constant that satifies intial condition}$$

$$r + 7 = 0 \,;\, r = -7$$

$$y = Ce^{-7t}$$

$$3 = Ce^{-7(0)}$$

$$C = 3 \,;\, y = 3e^{-7t}$$

7.) Find the value of the given interval.

$$\int_{\frac{\pi}{2}}^{\pi} 5\cos z \; dz$$

a. 5
b. 0
c. **-5**
d. 10

Correct Answer (C)

Solution: Referencing the Fundamentals of Engineering Supplied-Reference Handbook; Mathematics Section.

$$\int_{\frac{\pi}{2}}^{\pi} 5\cos(z) \; dz$$

$$5\left[\sin(z) \, \Big|_{\frac{\pi}{2}}^{\pi}\right]$$

$$5\left[\sin(\pi) - \sin\left(\frac{\pi}{2}\right)\right]$$

$$5[0-1] = -5$$

8.) What is the magnitude of the cross product to the following vectors;
A= 3i+2j and B= -i+2j-k.

   a. $\sqrt{24}$
   b. 8
   c. $\sqrt{77}$
   d. 9

Correct Answer (C)

Solution: Referencing the Fundamentals of Engineering Supplied-Reference Handbook; Mathematics Section.

The cross product of two vectors (A) and (B) is a vector perpendicular to (A) and (B).

$$\begin{bmatrix} i & j & k \\ 3 & 2 & 0 \\ -1 & 2 & -1 \end{bmatrix} = i(-2-0) - j(-3-0) + k(6-(-2))$$

$$= -2i + 3j + 8k$$

$$\text{Finding the magnitude: } \sqrt{(-2)^2 + (3)^2 + (8)^2}$$

$$= \sqrt{77}$$

9.) What is the equation of the circle passing through points (x,y) of (-4,0), (0,0), and (0,4).

   a. $(x+2)^2 + (y-3)^2 = 18$
   b. $(x+2)^2 + (y-2)^2 = 8$
   c. $(x-3)^2 + (y+3)^2 = 18$
   d. $(x+3)^2 + (y+9)^2 = 16$

Correct Answer (B)

Solution: Referencing the Fundamentals of Engineering Supplied-Reference Handbook; Mathematics Section.

The center-radius form for the equation of a circle is;

$$(x - h)^2 + (y - k)^2 = r^2$$

Substituting in points (0,0) and (0,4).

1. $(0 - h)^2 + (0 - k)^2 = r^2$
2. $(0 - h)^2 + (4 - k)^2 = r^2$

Since Equations 1 and 2 are both equal to r², setting them equal to one another and solving for k gives us;

$$h^2 + k^2 = -h^2 + (4 - k)^2$$
$$k^2 = (4 - k)^2$$
$$k = 2$$

Substituting the final point (-4,0) into the center-radius form equation gives us;

$$(-4 - h)^2 + (0 - k)^2 = r^2$$

Again, setting this equation equal to the equation for the point (0,0) and solving for h gives us;

$$(-4 - h)^2 + k^2 = h^2 + k^2$$
$$(-4 - h)^2 = h^2$$
$$h = -2$$

Now that h and k are known, substituting back into the equation for point (0,0) to solve for r gives us;

$$h^2 + k^2 = r^2$$
$$(-2)^2 + (2)^2 = r^2$$
$$r = \sqrt{8}$$

Substituting h, k, and r back into the center radius equation gives us;

$$(x + 2)^2 + (y - 2)^2 = \sqrt{8}^2 = \mathbf{8}$$

10.) What is the value of $\ln(e^{6.7})$?

    a. 2
    b. **6.7**
    c. 1.99
    d. 8.49

Correct answer (B)

Solution: Referencing the Fundamentals of Engineering Supplied-Reference Handbook; Mathematics Section.

Logarithm is shown as;

$$\log_b b^n = n$$

$$\ln(e^{6.7}) = \mathbf{6.7}$$

11.) From a group of 8 men and 4 women, a team of 4 will be formed by random selection. What is the probability that the team will consist of 4 men?

a. 0.50
b. 0.267
c. 0.048
d. **0.141**

Correct Answer (D)

Solution: Referencing the Fundamentals of Engineering Supplied-Reference Handbook; Mathematics Section; Statistics.

$$P(4\ Men) = \frac{\binom{8}{4} * \binom{4}{0}}{\binom{12}{4}} = \mathbf{0.141}$$

12.) The cycle times for an intermittent flare in a natural gas operation forms a normal distribution, with a standard deviation of 2 minutes. The mean cycle time is 11 minutes. The probability that a cycle selected at random will last more than 15 minutes is;

a. 0.0179
b. 0.9821
c. **0.0228**
d. 0.0540

Correct Answer (C)

Solution: Referencing the Fundamentals of Engineering Supplied-Reference Handbook; Mathematics Section; Statistics.

$$Z = \frac{Cycle\ time - \mu}{\sigma}$$

$$Z = \frac{15 - 11}{2} = 2\ Standard\ Deviations$$

Referencing the normal distribution table in the Fundamentals of Engineering Supplied-Reference Handbook; for x =2, **R(x)= 0.0228.**

13.) What is the standard deviation of these four given values;

       1,3,5,7

a. $\sqrt{5}$
b. 4
c. $\sqrt{6}$
d. 3.5

Correct Answer (A)

Solution: Referencing the Fundamentals of Engineering Supplied-Reference Handbook; Mathematics Section; Statistics.

$$\sigma = \sqrt{\left(\frac{1}{N}\right)\sum(X_i - \bar{x})^2}$$

$$\bar{x} = \frac{16}{4} = 4$$

X	$x - \bar{x}$	$(x - \bar{x})^2$
1	-3	9
3	-1	1
5	1	1
7	3	9
$\sum 16$		$\sum 20$

$$\sigma = \sqrt{\frac{20}{4}} = \sqrt{5}$$

14.) What is the approximate probability that no two people in a randomly selected group of six having the same birthday?

 a. 0.98
 b. 0.94
 c. 0.04
 d. **0.96**

Correct Answer (D)

Solution: Referencing the Fundamentals of Engineering Supplied-Reference Handbook; Mathematics Section; Statistics.

The probability of the first person having a unique birthday = $\frac{365}{365}$

2nd Person's Probability = $\frac{365-1}{365}$, 3rd Person's Probability = $\frac{365-2}{365}$,

4th Person's Probability = $\frac{365-3}{365}$, 5th Person's Probability = $\frac{(365-4)}{365}$,

6th Person's Probability = $\frac{365-5}{365}$; P(Six distinct Birthdays) = P(1)*P(2)*P(3)*P(4)*P(5)*P(6)

$$\frac{365}{365} * \frac{365-1}{365} * \frac{365-2}{365} * \frac{365-3}{365} * \frac{365-4}{365} * \frac{365-5}{365} = 0.9595$$

15.) A haul truck has an initial cost of $100,000, and a salvage value of $20,000 after 10 years of use. Assuming straight line deprecation, what is the value of the haul truck after six years?

   a. $60,000
   b. $48,000
   c. $8,000
   d. **$52,000**

Correct Answer (D)

Solution: Referencing the Fundamentals of Engineering Supplied-Reference Handbook; Engineering Economics Section; Depreciation.

Straight Line Depreciation Formula is shown as;

$$D_j = \frac{C - S_n}{n}$$

$$D_j = \frac{\$100,000 - \$20,000}{10} = \$8,000 \; per \; year$$

$$\$100,000 - \left(\frac{\$8,000}{year} * 6 \; Years\right) = \$52,000$$

16.) A distiller can produce a bottle of gin for $5 that can be sold for $15. The factory has operating and maintenance costs of $80,000 per year and $100,000 per year of capital costs. How many bottles of gin must the company produce and sell to break even on the year?

   a. **18,000**
   b. 21,000
   c. 300
   d. 45,000

Correct Answer (A)

Solution: Referencing the Fundamentals of Engineering Supplied-Reference Handbook; Engineering Economics Section; Break Even Analysis.

$$Revenue = \sum Costs$$

$$\$15x = \$5x + \$80,000 + \$100,000$$

$$x = \mathbf{18,000\ bottles}$$

17.) A crusher for use in a mining operation is expected to have a maintenance cost of $10,000 in the first year. The maintenance costs are expected to increase $1,000 per year. The interest rate is 8% per year, compounded annually. Over a five year period, what will be the approximate effective annual maintenance costs?

   a. $12,500
   b. $13,625
   c. **$11,850**
   d. $10,890

Correct Answer (C)

Referencing the Fundamentals of Engineering Supplied-Reference Handbook; Engineering Economics Section.

The Uniform Gradient Uniform Series is shown as;

$$A = A_1 + G\left(\frac{A}{G}, 8\%, 5\right)$$

$$\$10,000 + \$1,000(1.8465) = \mathbf{\$11,847}$$

18.) A county is expecting the need for a new wastewater treatment system that will cost $4 million in four years. If they were to invest funds today in an account earning 18% per year, how much would they need to invest?

   a. **$2,100,000**
   b. $6,100,000
   c. $2,750,000
   d. $3,200,000

Correct Answer (A)

Referencing the Fundamentals of Engineering Supplied-Reference Handbook; Engineering Economics Section.

The Single Payment Present Worth Formula is;

$$P = F\left(\frac{P}{F}, 18\%, 4\right)$$

$$P = \$4,000,000(0.5158) = \$2,063,200$$

19.) What is the value of 65 MPa in the following illustration called?

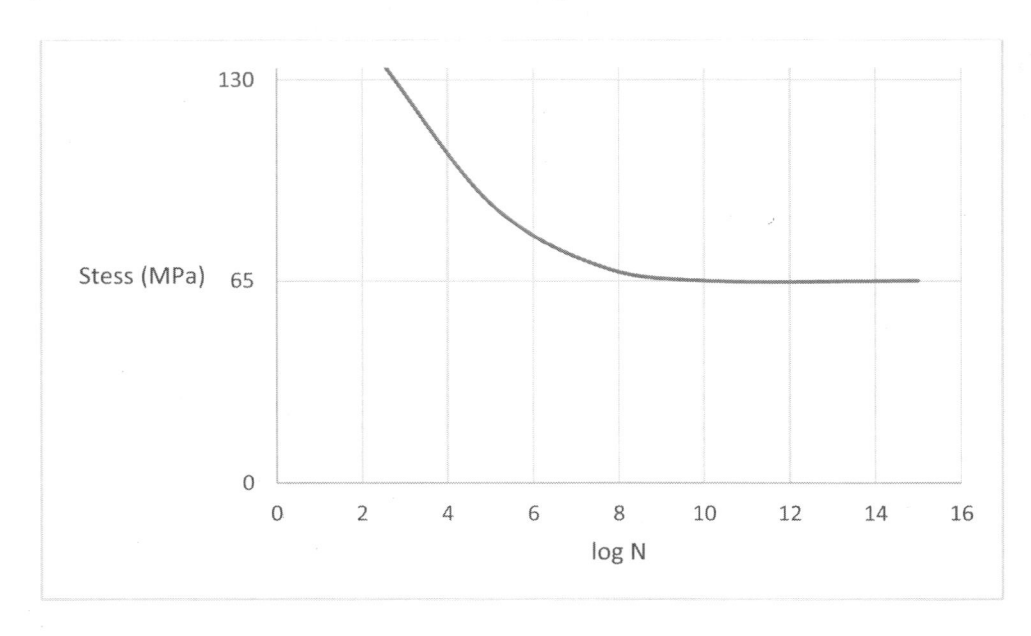

   a. **Fatigue Limit**
   b. Proportional Limit
   c. Yield Point

d. Tensile Strength

Correct Answer (A)

Referencing the Fundamentals of Engineering Supplied-Reference Handbook; Material Science; Structure of Matter Section.

This illustration details the results of an endurance test performed on a specific material. The value of 65 MPa is referred to as the fatigue limit, and represents the maximum stress that can be repeated without causing failure in the material.

20.) Galvanic action results from a difference in what property of metallic ions?

   a. Half-cell potential
   b. Specific heat
   c. Defect Flux
   d. **Oxidation Potential**

Correct Answer (D)

Referencing the Fundamentals of Engineering Supplied-Reference Handbook; Material Science; Structure of Matter Section.

Galvanic action results in corrosion. The greater the difference in oxidation potentials the greater the galvanic action, therefore the greater corrosion.

21.) While the corrosion of cast iron can be limited with a more electropositive coating, a coating that is less electropositive will lead to increased corrosion. Which of the following would be a poor choice to use as coating material on cast iron?

   a. Low carbon steel
   b. 2024 Aluminum Alloy
   c. Alcad 3S
   d. **Silver**

Correct Answer (D)

Solution: Referencing the Fundamentals of Engineering Supplied-Reference Handbook; Material Science; Structure of Matter Section.

Low carbon steel, 2024 Aluminum alloy, and Alcad 35 are all more electropositive that cast iron, so they would inhibit corrosion. Silver is less electropositive however, and will lead to more corrosion.

22.) A 1 Liter solution contains the following; 37 g H_2SO_4, 201 g $KMnO_4$, 13.8 g K_2SO_4, and 3.5 g Mn_2O_7. The equation for this reaction can be seen below. What is the equilibrium constant for this reaction?

$$H_2SO_4 + 2KMnO_4 \leftrightarrow K_2SO_4 + Mn_2O_7 + H_2O$$

a. 0.621
b. **0.002**
c. 1.0
d. 0.006

Correct Answer (B)

Solution: Referencing the Fundamentals of Engineering Supplied-Reference Handbook; Chemistry Section.

The equation for an Equilibrium Constant of a Chemical Reaction is shown as;

$$aA + bB \leftrightarrow cC + dD$$

$$K_{eq} = \frac{[C]^c[D]^d}{[A]^a[B]^b}$$

The first step is to figure out the Molarity of the different reactants.

$$[K_2SO_4] = \frac{\left(13.8\ g * \frac{1\ mole}{174\ g}\right)}{1L} = [0.079]$$

$$[Mn_2O_7] = \frac{\left(3.5\ g * \frac{1\ mole}{222\ g}\right)}{1L} = [0.016]$$

$$[H_2SO_4] = \frac{\left(37\ g * \frac{1\ mole}{98\ g}\right)}{1L} = [0.378]$$

$$[KMnO_4] = \frac{\left(201 \; g * \frac{1 \; mole}{158 \; g}\right)}{1L} = [1.272]$$

$$K_{eq} = \frac{[0.079]^1[0.016]^1}{[0.378]^1[1.272]^2} = 0.002$$

23.) What volume of 2 M HCl is required to neutralize 20 mL of 5 M NaOH?

a. 55 mL
b. 37mL
c. **50 mL**
d. 100 mL

Correct Answer (C)

Solution: Referencing the Fundamentals of Engineering Supplied-Reference Handbook; Chemistry Section.

The equation for this reaction is as follows;

$$HCl + NaOH \leftrightarrow H_2O + NaCl \qquad (1 \; mole \; of \; HCl \; neutralizes \; 1 \; mole \; of \; NaOH)$$

The first step is to figure out the total moles of NaOH in the 20 mL solution.

$$20 \; mL * \frac{L}{1000 \; mL} * \frac{5 \; mol}{L} = 0.1 \; mole \; NaOH$$

Therefore to neutralize the 0.1 mole of NaOH, 0.1 mole of HCl is necessary.

$$\frac{[0.1 \; mole \; HCl]}{[2 \; mole \; HCl/L]} = 0.05 \; L * 1000 \frac{mL}{L} = 50 \; mL$$

24.) What is the empirical formula for a compound containing 62.16% Iron (Fe), 35.61% Oxygen (O), and 2.23% Hydrogen (H) by percent?

a. Fe(OH)$_4$
b. Fe(OH)
c. **Fe$_2$(OH)**
d. Fe(OH)$_2$

Correct Answer (C)

Solution: Referencing the Fundamentals of Engineering Supplied-Reference Handbook; Chemistry Section; Stoichiometry.

The molecular weights for the elements are as follows: Fe-55.847 g/mole, O-16 g/mole, H-1 g/mole.

Dividing the atomic weights by the percent mass composition gives us:

$$Fe = \frac{55.847}{62.16} = 0.898 \qquad O = \frac{16}{35.61} = 0.449 \qquad H = \frac{1}{2.23} = 0.448$$

Since the ratio for Hydrogen is the smallest, we can divide the other ratios by the Hydrogen ratio.

$$Fe \approx \frac{0.898}{0.448} = 2 \qquad O \approx \frac{0.449}{0.448} = 1 \qquad H \approx \frac{0.448}{0.448} = 1$$

Therefore the equation is:

$$\boldsymbol{Fe_2(OH)}$$

25.) The oxidation of chromium in a ground water supply is determined to follow first order kinetics, with a rate constant of 0.10 per day. If the resulting concentration is 5.6 mg/L after 10 days, what was the initial concentration? Assume the ground water acts as an ideal plug flow reactor.

a. 32 mg/L
b. **15 mg/L**
c. 65 mg/L
d. 12 mg/L

Correct Answer (B)

Solution: Referencing the Fundamentals of Engineering Supplied-Reference Handbook; Environmental Engineering Section; Steady-State Reactor Parameters.

The equation for a first order ideal plug flow reactor is;

$$\theta = \frac{\ln\left(\frac{C_0}{C_t}\right)}{k}$$

Solving for C_0

$$C_0 = e^{\theta k} * C_t = e^{\frac{0.10}{day}*10\ day} * 5.6\frac{mg}{L} = 15\ mg/L$$

26.) The following chemical equilibrium constant equation shows the stoichiometry of which of the following reactions?

$$K_{EQ} = \frac{[C]^3[D]^6}{[A]^1[B]^2}$$

a. **A+2B ↔ 3C+6D**
b. 2A+3B↔ 3C+6D
c. 3C+6D↔ A+3B
d. A+B₂↔ C3+D6

Correct Answer (A)

Solution: Referencing the Fundamentals of Engineering Supplied-Reference Handbook; Chemistry Section.

The equation for an Equilibrium Constant of a Chemical Reaction is shown as;

$$aA + bB \leftrightarrow cC + dD$$

$$K_{eq} = \frac{[C]^c[D]^d}{[A]^a[B]^b}$$

27.) Hydrogen Bromide (HBr) is neutralized with barium hydroxide by the equation seen below. If you have 25 mL of 0.1 M HBr, how much 0.05 M Ba(OH)$_2$ solution will be needed for neutralization by the reaction shown below?

$$2HBr + Ba(OH)_2 \leftrightarrow BaBr_2 + 2H_2O$$

a. 30 mL
b. 20 mL
c. 25L
d. **25 mL**

Correct Answer (D)

Solution: Referencing the Fundamentals of Engineering Supplied-Reference Handbook; Chemistry Section.

The first step is to find the number of moles of HBr that need to be neutralized.

$$0.1 \frac{mole\ HBr}{L} * 25mL * \frac{L}{1000\ mL} = 0.0025\ mole\ HBr$$

Since the reaction given above shows that 1 mole of Ba(OH)$_2$ neutralizes 2 moles of HBr;

$$0.0025\ mole\ HBr * \frac{1 mole\ Ba(OH)_2}{2\ mole\ HBr} = 0.00125\ mole\ Ba(OH)_2$$

$$\frac{0.00125\ mole\ Ba(OH)_2}{0.05\ mole\ Ba(OH)_2/L} * \frac{1000mL}{L} = 25\ mL$$

28.) What coefficients (a,b,c,d) balance the following equation?

$$aC_9H_{10}O + bO_2 \leftrightarrow cCO_2 + dH_2O$$

a. 1,14,9,10
b. **1,11,9,5**
c. 2,23,19,10
d. 3,9,8,6

Correct Answer (B)

Solution: Referencing the Fundamentals of Engineering Supplied-Reference Handbook; Chemistry Section.

In this combustion reaction oxygen is added to a carbon chain, so the first step of the problem is to balance the carbon;

$$C_9H_{10}O + O_2 \leftrightarrow CO_2 + H_2O$$

$9a\ carbons = 1c\ Carbon;\ To\ balance\ the\ carbons\ the\ ratio\ a = 9c\ must\ be\ valid$

Next balance the hydrogen.

$$C_9H_{10}O + O_2 \leftrightarrow CO_2 + H_2O$$

$10a\ H = 2d\ H; To\ balance\ the\ hydrogens\ the\ ratio\ a = 5d\ must\ be\ valid$

Finally balance the oxygen.

$$C_9H_{10}O + O_2 \leftrightarrow CO_2 + H_2O$$

$$1 + 2(b) = 18 + 5$$

Solving for b gives us this equation;

$$C_9H_{10}O + 11O_2 \leftrightarrow 9CO_2 + 5H_2O$$

29.) Name the following compound;

$$CH_3 - CH_2 - CH_2 - \underset{\underset{CH_3}{|}}{\overset{\overset{CH_3}{|}}{C}} - CH = CH_2$$

a. 2,2-diethylpent-1-ene
b. 2-methylhex-2-ene
c. 3,3-diethylhex-2-ene
d. **3,3-dimethylhex-1-ene**

Correct Answer (D)

Solution: Referencing the Fundamentals of Engineering Supplied-Reference Handbook; Chemistry Section; Organic Chemistry.

Start by naming the carbon chain, which has six carbons so is "hexane". Then account for the two methyl groups that are at 3 and 3 giving us 3,3-dimethylhexane. Finally account for the double bond at the first carbon group, giving us **3,3-dimethylhex-ene.**

30.) Name the following compound;

$$CH_3 - CH_2 - CH_2 - CH = CH - CH_3$$

a. **Hex-2-ene**
b. Hep-2-ene
c. Hex-4-ene
d. Pent-2-ene

Correct Answer (A)

Solution: Referencing the Fundamentals of Engineering Supplied-Reference Handbook; Chemistry Section; Organic Chemistry.

Name the compound from the side which gives the smallest numbers to describe. The six carbons in the compound would make it a "hexane". To account for the double bond at the second carbon chain the -2-ene must be placed in the naming, giving you **Hex-2-ene**.

31.) A general approach to organic chemistry reactions assumes that the characteristic properties of compounds are determined by what?

   a. Alkalinity
   b. Structure of the compound
   c. Alcoholic content
   d. **Functional groups**

Correct Answer (D)

Solution: Referencing the Fundamentals of Engineering Supplied-Reference Handbook; Chemistry Section; Organic Chemistry.

The functional groups within a compound determine the characteristic properties of the compounds such as melting point, boiling point, and solubility.

32.) A stream has an initial BOD concentration of 500 mg/L and an initial dissolved oxygen deficit in the mixing zone of 2.2 mg/L. Assuming the reaeration of the water follows a rate constant of 0.6/day and the deoxygenation occurs at a rate of 0.7/ day, what is the dissolved oxygen concentration after 10 days according to the Steeter Phelps equation? The saturated dissolved oxygen concentration is 10 mg/L.

   a. 7.8 mg/L
   b. **4.5 mg/L**
   c. 10 mg/L
   d. 5.5 mg/L

Correct Answer (B)

Solution: Referencing the Fundamentals of Engineering Supplied-Reference Handbook; Environmental Engineering Section; Stream Modeling-Streeter Phelps Equation.

The Streeter Phelps Equation is shown as;

$$D = \frac{k_1 L_0}{k_2 - k_1} [\exp(-k_1 t) - \exp(-k_2 t)] + D_0 \exp(-k_2 t)$$

The following variables are given in the problem;

$$t = 10\ days; \quad k_1 = \frac{0.7}{day} \quad k_2 = \frac{0.6}{day} \quad D_0 = 2.2\frac{mg}{L} \quad L_0 = 500\ mg/L$$

Inserting those values into the equation gives us,

$$D = \frac{(\frac{0.7}{day} * 500\frac{mg}{L})}{(\frac{0.6}{day} - \frac{0.7}{day})}\left[\exp(-\frac{0.7}{day} * 10\ day) - \exp\left(\frac{-0.6}{day} * 10\ day\right)\right]$$
$$+ 2.2\frac{mg}{L}\exp\left(-\frac{0.6}{day} * 10\ day\right)$$

$$= 5.49\frac{mg}{L}$$

$$Since\ DO = DO_{SAT} - D$$

$$DO = 10\frac{mg}{L} - 5.49\frac{mg}{L} = \mathbf{4.5\ mg/L}$$

33.) How many grams of carbon dioxide is dissolved in 1 Liter of carbonated water if the manufacturer uses a pressure of 2 atm in the bottling process at 25°C? Note that the K_H of CO_2 in water is equal to 29.76 atm/(mol/L) at 25°C.

   a. 4.5 g
   b. **3.0 g**
   c. 4.5 g
   d. 12 g

Correct Answer (B)

Solution: Referencing the Fundamentals of Engineering Supplied-Reference Handbook; Environmental Engineering Section.

Henry Law is as follows;

$$P_i = K_H C$$

$$C = \frac{P_i}{K_H}$$

$$C = \frac{2 \; atm}{29.76 \frac{atm}{\frac{mole}{L}}} = 0.0672 \frac{mol}{L}$$

$$\left(0.0672 \frac{mole}{L}\right)(1 \; L)\left(44 \frac{g}{mole \; CO_2}\right) = 2.96 \; g$$

34.) A body of water that sees a large increase in the concentration of phosphates and nitrates from an influent waste stream will most likely be what?

a. **Eutrophic**
b. Oligotrophic
c. Mesotrophic
d. Nutrient Deficient

Correct Answer (A)

Solution: Referencing the Fundamentals of Engineering Supplied-Reference Handbook; Environmental Engineering Section.

Eutrophication is the enrichment of an ecosystem with chemical nutrients, typically compounds containing nitrogen, phosphorus, or both. A body of water that is receiving a large inflow of nitrates and phosphates would most likely be Eutrophic.

35.) What is the oxidation state of Chromium (Cr) in the following compound?

$$CrO_6^{-4}$$

a. 4
b. 7
c. **8**
d. 6

Correct Answer (C)

Solution: Referencing the Fundamentals of Engineering Supplied-Reference Handbook; Chemistry Section; Oxidation States.

Since the oxygen has an oxidation state of -2, and the compound has an overall oxidation state of -4;

$$(-2) * 6 + x = -4$$

$$x = 8$$

36.) In a skin exposure toxicology study, the median single dose that is expected to kill 50 percent of a group of test animals is known as which of the following?

a. LC_{50}
b. Dosage
c. **LD_{50}**
d. Concentration

Correct Answer (C)

Solution: Referencing the Fundamentals of Engineering Supplied-Reference Handbook; Environmental Engineering Section; Dose Response Curve.

The LD_{50} is the median single dose which is expected to kill 50% of a group of test animals usually by oral or skin exposure. The LC_{50} is the median air concentration that is expected to kill 50% of a group of test animals when administered as a single exposure over one or four hours.

37.) A 70 kg man consumes 80 mg/day of soil that contains 20 mg/kg of benzene. He consumes this soil every day for 70 years. If the cancer slope factor for benzene is 0.029 kg*d/mg and the absorption factor is 1, his risk of developing cancer is most near which of the following?

a. 9.1×10^{-7}
b. 4.8×10^{-5}
c. 2.3×10^{-5}
d. **6.6×10^{-7}**

Correct Answer (D)

Solution: Referencing the Fundamentals of Engineering Supplied-Reference Handbook; Environmental Engineering Section.

The exposure equation for ingestion of chemicals is soil is as follows;

$$CDI = \frac{(CS)(IR)(CF)(FI)(EF)(ED)}{(BW)(AT)}$$

$$CDI = \frac{\left(20\frac{mg}{kg}\right)\left(80\frac{mg}{day}\right)\left(\frac{10^{-6}\,kg}{mg}\right)(70\,year)\left(365\frac{day}{year}\right)}{(70\,kg)(70\,year)\left(365\frac{day}{year}\right)} = 2.286 \times 10^{-5}\frac{mg}{kg*day}$$

$$Risk = CDI * SF = 2.286 \times 10^{-5}\frac{mg}{kg*day} * 0.029\frac{kg*day}{mg} = 6.63 * 10^{-7}$$

38.) A 40 hour per week worker in an industrial facility is known to be exposed to a compound in the air at a concentration of 80 µg/m³. The worker inhales at a rate of 1 m³ per hour, and has a body mass of 70 kg. The chronic daily intake of this worker during an eight hour shift is which of the following? Assume the fraction of the chemical that is absorbed in the lungs is 0.65.

a. **0.006 mg/(kg*day)**
b. 0.46 mg/(kg*day)
c. 0.04 mg/(kg*day)
d. 0.10 mg/(kg*day)

Correct Answer (A)

Solution: Referencing the Fundamentals of Engineering Supplied-Reference Handbook; Environmental Engineering Section; Exposure.

The first step is to determine the intake amount in one 8 hour work day;

$$Intake = 80\frac{\mu g}{m^3} * 1\frac{m^3}{hr} * \frac{10^{-3}mg}{\mu g} * 8\frac{hour}{day} * 0.65 = 0.416\frac{mg}{day}$$

$$CDI = \frac{0.416\frac{mg}{day}}{70\,kg} = \mathbf{0.006\frac{mg}{(kg*day)}}$$

39.) During a study of dose and response it is determined that the safe human dose of a chemical is 500 mg/day and the no-observed-adverse-effect-level ("NOAEL") for the chemical is 15 mg/kg*day. What uncertainty factor was used in the study, assuming it was conducted on average adult males?

a. **2.3**
b. 18.2
c. 4.2
d. 6.4

Correct Answer (A)

Solution: Referencing the Fundamentals of Engineering Supplied-Reference Handbook; Environmental Engineering Section; Reference Dose.

The equation for the Safe Human Dose (SHD) is shown as the following;

$$SHD = RfD * W = \frac{NOAEL * W}{UF} \qquad UF = \frac{NOAEL * W}{SHD}$$

$$UF = \frac{\left(15 \frac{mg}{kg * day}\right)(78 \; kg)}{\left(500 \frac{mg}{day}\right)} = 2.34$$

40.) For carcinogens, the general acceptable value for risk in a population is less than which of the following?

a. 10^{-2}
b. **10^{-6}**
c. 10^{-1}
d. 10^{-3}

Correct Answer (B)

Solution: Referencing the Fundamentals of Engineering Supplied-Reference Handbook; Environmental Engineering Section; Carcinogens.

The general acceptable value for risk for carcinogens is 1 per million, or 10^{-6}.

41.) A man has been drinking water from the same rural well for 70 years. After analytical sampling takes place it is determined that the water contains arsenic at concentrations of 5 mg/L. If the cancer slope factor of arsenic is 0.67 kg*day/mg, what is his cancer risk from this exposure? Assume an ingestion rate of 2 liters per day.

a. 0.15
b. **0.09**
c. 0.02
d. 1 X 10^{-6}

Correct Answer (B)

Solution: Referencing the Fundamentals of Engineering Supplied-Reference Handbook; Environmental Engineering Section; Exposure.

The equation for ingestion in drinking water is as follows;

$$CDI = \frac{(CW)(IR)(EF)(ED)}{(BW)(AT)}$$

Substituting in the known values;

$$CDI = \frac{(5\frac{mg}{L})(2\frac{L}{day})(365\frac{day}{year})(70\ year)}{(78\ kg)(70\ year * 365\frac{day}{year})} = 0.128\frac{mg}{kg * day}$$

Multiplying the Chronic Daily Intake by the cancer slope factor gives us the Cancer Risk.

$$0.128\frac{mg}{kg * day} * 0.67\frac{kg * day}{mg} = \mathbf{0.0859}$$

42.) Which of the following signifies the dosage below which there are no harmful effects?

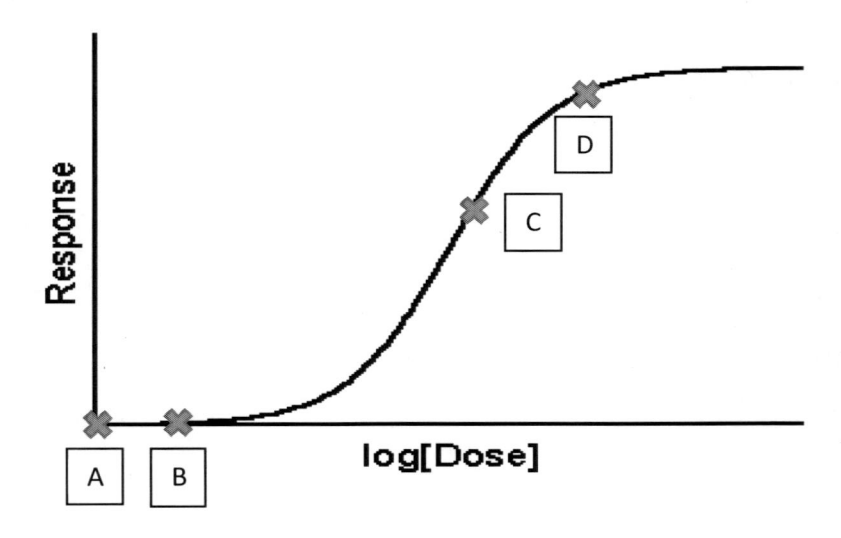

[Graph: Response vs log[Dose], sigmoid curve with points A, B on baseline and C, D on the rising/upper portion]

a. A
b. **B**
c. C
d. D

Correct Answer (B)

Solution: Referencing the Fundamentals of Engineering Supplied-Reference Handbook; Environmental Engineering Section; Non-Carcinogen Dose Response Curve.

Point B represent from the graph the point at which below the dose no response is observed in test subjects.

43.) A fluid with a vapor pressure of 0.4 Pa and a specific gravity of 10 is used in a barometer. If the fluids column height is 1 m, what is the atmospheric pressure?

a. **98 KPa**
b. 114 Kpa
c. 76 KPa
d. 84 KPa

Correct Answer (A)

Solution: Referencing the Fundamentals of Engineering Supplied-Reference Handbook; Fluid Mechanics Section.

The equation for a barometer is shown as;

$$P_{atm} = P_v + \rho g h$$

Substituting in known variables and assuming the density of water is 1000 kg/m³;

$$P_{atm} = \left[0.4\ Pa + 10(1000\tfrac{kg}{m^3})(9.81\tfrac{m}{s^2})(1\ m)\right] * \frac{KPa}{1000\ Pa} = \mathbf{98\ KPa}$$

44.) A pressure of 1000 lbf/ft² is measured 10 feet below the surface of an unknown liquid. What is the specific gravity of the liquid?

a. **1.60**
b. 1.25
c. 30.3
d. 0.35

Correct Answer (A)

Solution: Referencing the Fundamentals of Engineering Supplied-Reference Handbook; Fluid Mechanics Section.

Using the given pressure and height, the specific gravity can be calculated as follows;

$$\Delta p = \rho g \Delta h = (SG)\rho_{water} g \Delta h$$

$$SG = \frac{p}{\rho_{water} g \Delta h * 1/g_c}$$

$$\frac{1000\ \tfrac{lbf}{ft^2}}{62.4\ \tfrac{lbm}{ft^3} * \tfrac{32.2 ft}{sec^2} * 10\ ft * \tfrac{lbf * sec^2}{32.2\ lbm * ft}} = 1.6$$

45.) A six inch plastic pipe that is 1200 feet long has a Hazen Williams coefficient of 120. If the pipe is carrying 0.6 cfs of water and the pipe is full, the head loss in feet of water is most nearly?

a. 0.8
b. 2.5
c. 9
d. 15

Correct Answer (C)

Solution: Referencing the Fundamentals of Engineering Supplied-Reference Handbook; Fluid Mechanics Section.

The Hazen Williams Equation is shown as;

$$v = k_1 C R_H^{0.63} S^{0.54}$$

The first step is to determine the area of the pipe, then the velocity of the water.

$$A = \frac{\pi}{4}(D^2) = \frac{\pi}{4} * (0.5 \, ft)^2 = 0.1964 \, ft^2$$

$$V = \frac{Q}{A} = \frac{0.6 \frac{ft^3}{s}}{0.1964 \, ft^2} = 3.056 \frac{ft}{s}$$

Substituting the known values back into the Hazen Williams equation (k= 1.318 for USCS units) and solving for S gives us;

$$S^{0.54} = \frac{V}{1.318 C R_H^{0.63}} = \frac{3.056 \frac{ft}{s}}{\left[1.318 * 120 * \left(\frac{0.5}{4}\right)^{0.63}\right]} = 0.07161$$

$$0.07161^{\left(\frac{1}{0.54}\right)} = 0.00758$$

$$h_F = Length * Slope = 0.00758 * 1200 \, ft = \mathbf{9.1 \, ft \, water}$$

46.) A sanitary sewer system with a depth to diameter ratio of 0.6 has a flow capacity of 50 cfs. Assuming the roughness coefficient varies with depth, what is the designed flow rate for this system?

   a. 12 cfs
   b. 50 cfs
   c. 46 cfs
   d. **28 cfs**

Correct Answer (D)

Solution: Referencing the Fundamentals of Engineering Supplied-Reference Handbook; Civil Engineering Section; Hydraulic Elements Graph for Circular Sewers.

The Hydraulic Elements Graph for Circular Sewers shows that for a depth to diameter ration (d/D) of 0.6 the ratio of the flow rate to the full flow rate (Q/Q$_{FULL}$) is equal to 0.55. Therefore;

$$Q = Q_{FULL} * \frac{Q}{Q_{FULL}} = 50 \, cfs * 0.55 = \mathbf{28 \, cfs}$$

47.) A sanitary sewer system is being designed for an apartment complex. The calculations show a flow rate of 1.57 m³/s, a Manning's roughness coefficient of 0.014 and a diameter of 1 meter. What slope percent was used in the design?

   a. 0.25%
   b. 2.5%
   c. 0.05%
   d. **0.5%**

Correct Answer (D)

Solution: Referencing the Fundamentals of Engineering Supplied-Reference Handbook; Fluid Dynamics Section.

Manning's Equation is shown as:

$$Q = \frac{K}{n} A R_H^{2/3} S^{1/2}$$

$$S = \left[\frac{Qn}{KAR_H^{\frac{2}{3}}}\right]^2$$

$$\left[\frac{(1.57 \, m^3/sec)(0.014)}{1 * (1 \, m^2 * \frac{\pi}{4})(\frac{1}{4})^{2/3}}\right]^2 = 0.005 * 100 \, (percent) = 0.5\%$$

48.) The critical depth of a rectangular stream with a cross sectional area of 7 m² and a width of 3 m is most nearly?

a. **2.33 m**
b. 1.76 m
c. 1.54 m
d. 2.48 m

Correct Answer (A)

Solution: Referencing the Fundamentals of Engineering Supplied-Reference Handbook; Civil Engineering Section; Environmental Engineering Section.

The critical depth equation is defined as;

$$\frac{Q^2}{g} = \frac{A^3}{T}$$

$$Q = \sqrt{\frac{(7m^2)^3}{3m} * 9.81 \, m/s^2} = 33.5 \, m^3/sec$$

For rectangular channels the equation for critical depth is;

$$y_c = \left[\frac{q^2}{g}\right]^{1/3} \quad \text{where } q = Q/B$$

$$y_c = \left[\frac{\left(\frac{33.5 \, m^3}{s}\middle/3 \, m\right)^2}{9.81 \, \frac{m^3}{s}}\right]^{1/3} = 2.33 \, m$$

49.) Five cubic feet per second of water is being pumped through 1000 feet of eight inch pipe. Neglecting minor losses, the pump will need to be able to handle how many feet of total head? The static head is 20 feet, and the friction coefficient is 140.

a. **91 feet**
b. 83 feet
c. 74 feet
d. 45 feet

Correct Answer (A)

Solution: Referencing the Fundamentals of Engineering Supplied-Reference Handbook; Fluid Mechanics Section.

The total head is equal to the static head plus the frictional head (h_f). The static head is given as 20 feet. In order to determine the frictional head the slope of the pipe first needs to be known. Using the Hazen Williams Equation the slope can be found;

$$V = k_1 C R_H^{0.63} S^{0.54}$$

Solving for S gives us;

$$\left(\frac{V}{k_1 C R_H^{0.63}}\right)^{\left(\frac{1}{0.54}\right)} = S$$

$$V = \frac{Q}{A} = \frac{5 \, ft^3/s}{\left(\frac{8}{12}\right)^2 * \frac{\pi}{4}} = 14.32 \frac{ft^3}{s}$$

$$\left(\frac{\frac{14.32 \, ft^3}{s}}{1.318 * 140 * \left[\frac{\left(\frac{8}{12}\right)}{4}\right]^{0.63}}\right)^{\left(\frac{1}{0.54}\right)} = 0.071$$

$$h_f = S * L = 0.071 * 1000 = 71 \, ft$$

$$h_f + h_s = 71 \, feet + 20 \, feet = \mathbf{91 \, feet}$$

50.) A pump station includes two centrifugal pumps in parallel. The pump curve for a single pump can been seen below.

$$Head = 70 - 0.002Q^2$$
Where Q is in units of gpm, and the equation is valid from $50 \, gpm < Q < 300 \, gpm$

The piping system needs to overcome a static head of 40 feet. The friction loss for the system is defined by the following equation;

$$Friction \, Loss \, (feet) = 0.0004Q^2$$
Where Q is in units of gpm, and the equation is valid from $50 \, gpm < Q < 300 \, gpm$

When the system is operating a single pump, the flow rate is closest to which value?

a. 122 gpm
b. 150 gpm
c. **112 gpm**
d. 130 gpm

Correct Answer (C)

The curve for the system is equal to the static head plus the dynamic head.

$$System\ Curve = 40 + 0.0004Q^2$$

$$Pump\ Curve = 70 - 0.002Q^2$$

The point of operation is equal to the intersection point of the two curves

$$70 - 0.002Q^2 = 40 + 0.0004Q^2$$

$$Solving\ for\ Q = \left[\frac{70-40}{0.0004+0.002}\right]^{\frac{1}{2}} = 112\ gpm$$

51.) A rectangular suppressed weir is being sized to handle a flow rate of 3 cubic meters per second. If the depth of the water in the weir needs to be less than 1 meter, how long does the weir need to be?

a. 1.8 m
b. **1.6 m**
c. 3.2 m
d. 1 m

Correct Answer (B)

Solution: Referencing the Fundamentals of Engineering Supplied-Reference Handbook; Civil Engineering Section.

The weir formula for free discharge suppressed is shown as;

$$Q = CLH^{3/2}$$

Solving for L gives us;

$$\frac{Q}{H^{\frac{3}{2}}C} = L = \frac{3\ m^3/s}{(1m)^{3/2} * 1.84} = 1.63\ m$$

52.) The orifice shown below is discharging freely into the atmosphere.

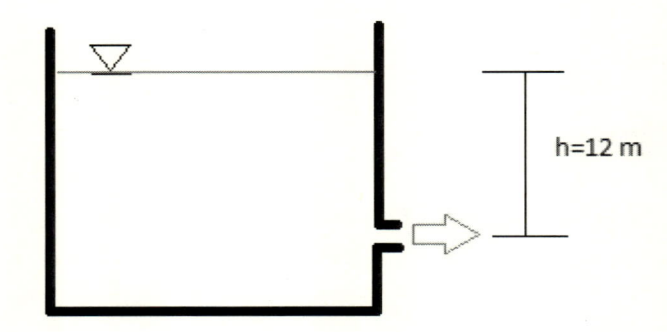

If the coefficient of discharge is 0.34 and the diameter of the discharge pipe is 0.25m, what is the resulting flow rate?

e. 0.20 m³/sec
f. 0.15 m³/sec
g. 0.12 m³/sec
h. **0.26 m³/sec**

Correct Answer (D)

Solution: Referencing the Fundamentals of Engineering Supplied-Reference Handbook; Fluid Mechanics Section.

The equation for an Orifice discharging freely into the Atmosphere is shown as;

$$Q = CA_0\sqrt{2gh}$$

$$Q = 0.34 * (0.25\ m)^2 * \frac{\pi}{4} * \sqrt{2 * 9.81 \frac{m}{s^2} * 12\ m} = 0.26\ \frac{m^3}{s}$$

53.) A hose shoots out a jet of water vertically with a velocity (v) and a flow rate (Q). A horizontal plate is located is located directly above the nozzle at a height (h). The density of the water is ρ. What is the force necessary to keep the plate in equilibrium against the force of the water jet?

    a. $Q\rho\sqrt{v^2 - 2gh}$
    b. $\sqrt{v^2 2gh}$
    c. $Q\rho\sqrt{v - 2gh}$
    d. $\frac{1}{2}mv_f^2$

Correct Answer (A)

Solution: Referencing the Fundamentals of Engineering Supplied-Reference Handbook; Fluid Dynamics Section.

The first step is to find the velocity of the water as it impacts the plate. This can be done by using the conservation of energy principle (found in the Dynamics Section). Note that when the water leaves the hose it has only kinetic energy, and when it impacts the plate it has both kinetic and potential energy. Therefore the following is true for the plate to be in equilibrium;

$$\frac{1}{2}mv^2 = \frac{1}{2}mv_f^2 + mgh$$

The velocity of the water can then be solved for at the equilibrium conditions;

$$v_f = \sqrt{v^2 - 2gh}$$

The force of the jet on the plate is then;

$$F = Q\rho v_f = Q\rho\sqrt{v^2 - 2gh}$$

54.) An oil pipeline pumps 1000 barrels of crude oil per minute with a specific gravity of 0.68. The static head on the system is 40 feet. Neglecting minor losses, how much power must be applied by a pump assuming 100% efficiency?

a. 425 hp
b. **290 hp**
c. 300 hp
d. 255 hp

Correct Answer (B)

Solution: Referencing the Fundamentals of Engineering Supplied-Reference Handbook; Fluid Mechanics Section.

The pump power equation is shown as;

$$\dot{W} = Q\gamma h_p$$

$$Q = \left[1000 \frac{barrels}{min} * 42 \frac{gallons}{barrel} * \frac{ft^3}{7.48 \; gallon} * \frac{min}{60 \; sec}\right] = 93.58 \frac{ft^3}{sec}$$

$$\dot{W} = \left[93.58 \frac{ft^3}{sec} * 62.4 \frac{lbf}{ft^3} * 0.68 * 40 \; ft\right] = 158{,}836 \frac{lbf * ft}{sec}$$

Converting to Horsepower:

$$158{,}836 \frac{lbf * ft}{sec} * \frac{hp}{550 \frac{lbf * ft}{sec}} = \mathbf{289 \; hp}$$

55.) Air is compressed in a system to 1/5 of its initial volume. The final temperature is 500°C, and the process is frictionless and adiabatic. What is the initial temperature?

a. **138°C**
b. 124°C
c. 102°C

d. 94°C

Correct Answer (A)

Solution: Referencing the Fundamentals of Engineering Supplied-Reference Handbook; Thermodynamics Section.

The equation for a constant entropy (adiabatic) process is show as;

$$\frac{T_2}{T_1} = \left(\frac{V_1}{V_2}\right)^{k-1} \quad where\ k = \frac{c_p}{c_v}$$

Solving for T₁ gives us;

$$T_1 = \frac{T_2}{\left(\frac{V_1}{V_2}\right)^{k-1}} = \frac{[500 + 273\ K]}{[5]^{\frac{1000}{718}-1}} = 411\ K$$

Convert back to Celsius

$$411K - 273 = \mathbf{138°C}$$

56.) A refrigeration system is said to operate in a Carnot cycle. The system receives heat from a reservoir at 0°C. If the coefficient of performance ("COP") for the system is 3, what is the power input per ton of refrigeration?

a. 1355 W/ton
b. 11,500 W/ton
c. 1285 W/ton
d. **1180 W/ton**

Correct Answer (D)

Solution: Referencing the Fundamentals of Engineering Supplied-Reference Handbook; Thermodynamics Section; Basic Cycles.

The equation for a refrigeration cycle COP in refrigerators and air conditioners is shown as;

$$COP = \frac{Q_L}{W} = \frac{\dot{Q}_L}{\dot{W}}$$

Recognizing that one ton of refrigerant corresponds to 3516 W and solving for P;

$$\frac{\dot{Q}_L}{COP} = P = \frac{3516 \frac{W}{ton}}{3.00} = 1172 \frac{W}{ton}$$

57.) An air sample taken has a temperature of 25°C and a relative humidity of 30%. What temperature is the dew point closest to?

a. 10°C
b. 15°C
c. **5°C**
d. 0.01°C

Correct Answer (C)

Solution: Referencing the Fundamentals of Engineering Supplied-Reference Handbook; Thermodynamics Section.

The dew point temperature is the saturation temperature for the current vapor pressure conditions.

$$\emptyset = \frac{p_v}{p_g} = 0.3$$

From the Steam Tables it is found at 25°C the Saturation pressure for water is 3.169 KPa.

$$p_v = \emptyset p_g = 0.3 * 3.169 \, KPa = 0.9507 \, KPa$$

From the Steam Tables for 0.9507 KPa the dew point is closest to **5°C**.

58.) What are the products of complete combustion of gaseous hydrocarbons?

   a. Only Carbon Monoxide
   b. Only Carbon Dioxide
   c. Carbon Dioxide, Carbon Monoxide, and Water
   d. **Carbon Dioxide and Water**

Correct Answer (D)

Solution: Referencing the Fundamentals of Engineering Supplied-Reference Handbook; Thermodynamics Section.

Complete combustion of hydrocarbons produces only carbon dioxide and water. Carbon monoxide is formed by incomplete combustion of hydrocarbons.

59.) A city with a population of 40,000 people has an average water usage of 180 gallons/person per day. If the return rate is 75%, what is the maximum daily flow rate for wastewater?

   a. 5.4 MGD
   b. **14.6 MGD**
   c. 2.2 MGD
   d. 11 MGD

Correct Answer (B)

Solution: Referencing the Fundamentals of Engineering Supplied-Reference Handbook; Civil Engineering Section; Environmental Engineering Section.

The first step is to find the daily flow rate;

$$40,000 \; people * 180 \frac{gallons}{person * day} * 75\% = 5.4 \; MGD$$

Then, use the sewage flowrate curves to obtain a maximum flow factor (since the curve to use is not specified, use an average value of curve A and C).

$$Curve \; A = 2.4 \qquad Curve \; C = 3 \qquad AVG = 2.7$$

$$Q_{MAX\ DAILY} = 2.7 * 5.4\ MGD = \mathbf{14.6\ MGD}$$

60.) A small town is growing at an exponential rate of 2.2% per year. If the per capita water usage is 150 gallons per day, and the town is planning a drinking water plant to last for 20 years, what capacity should the plant be designed for? The current population is 5,500.

   a. **1.3 MGD**
   b. 0.8 MGD
   c. 0.5 MGD
   d. 1.8 MGD

Correct Answer (A)

Solution: Referencing the Fundamentals of Engineering Supplied-Reference Handbook; Environmental Engineering Section.

The equation for exponential population equation is shown as;

$$P_t = P_0 e^{k\Delta t} = 5,500 * e^{(0.022*20)}$$

$$P_{20} = 8,540\ people * 150\frac{gallons}{day * person} * \frac{MGD}{\frac{10^6 gallons}{day}} = \mathbf{1.3\ MGD}$$

61.) A metropolitan area is growing at a rate of 3% per year. If this is assumed to be a linear growth rate, what will be the population in 3 years if the current population is 1.2 million?

   a. 1.4 million
   b. 1.7 million
   c. **1.3 million**
   d. 1.9 million

Correct Answer (C)

Solution: Referencing the Fundamentals of Engineering Supplied-Reference Handbook; Environmental Engineering Section.

The formula for Linear Population Projection is shown as;

$$P_t = P_0 + k\Delta t$$

$$k = 1{,}200{,}000 * 0.03 = \frac{36{,}000}{year}$$

$$P_3 = 1{,}200{,}000 + \frac{36{,}000}{year} * 3 \ year = \mathbf{1{,}308{,}000 \ people}$$

62.) A storm produced 2 inches of water in 30 minutes. What is the probability of a storm of this intensity occurring during a given year according to the following graph?

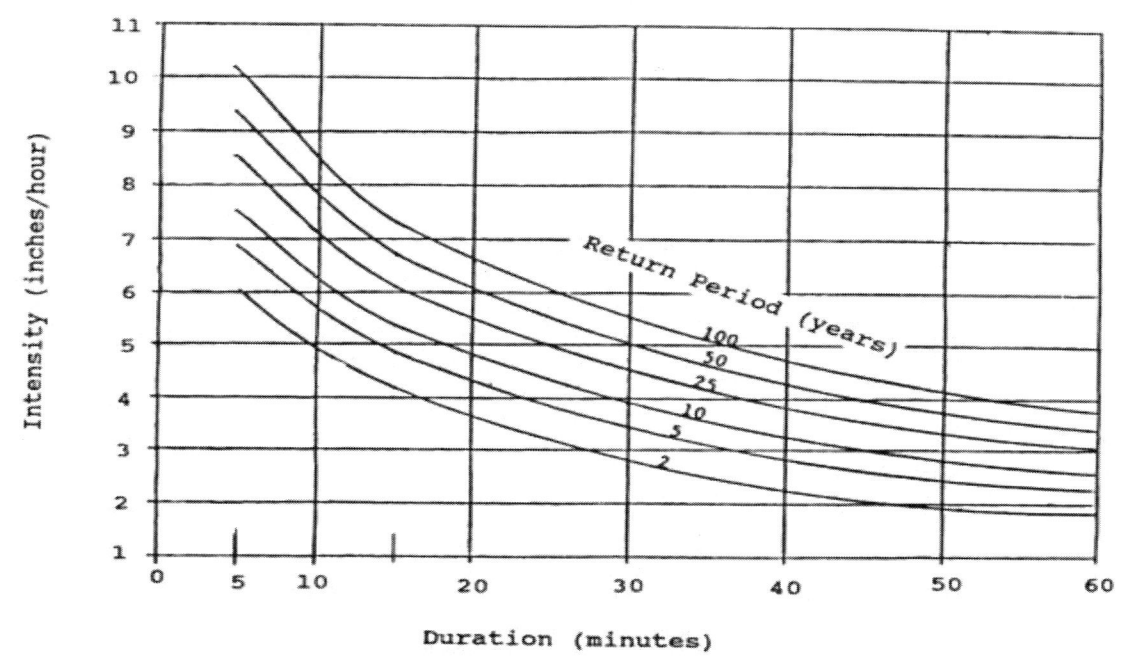

a. **0.10**
b. 0.50
c. 0.02
d. 0.01

Correct Answer (A)

Solution: Referencing the above graph and the Fundamentals of Engineering Supplied-Reference Handbook; Environmental Engineering Section.

A 30 minute storm with 2 inches of rainfall has an intensity of 4 inches per hour and intersects the 10 year return period curve, resulting in a probability of 0.10.

63.) A watershed with an area of 160 acres used to be pasture land with a runoff coefficient of 0.15. The area was developed into a subdivision with a runoff coefficient of 0.35. During similar rain events, an increase in runoff of 96 cfs was measured in the developed area. What was the intensity of the rainfall event?

   a. 2 in/hr
   b. **3 in/hr**
   c. 4 in/hr
   d. 5 in/hr

Correct Answer (B)

Solution: Referencing the Fundamentals of Engineering Supplied-Reference Handbook; Civil Engineering Section; Environmental Engineering Section.

The rational equation for runoff is shown as;

$$Q = CIA$$

$$Q_2 - Q_1 = 96 \; cfs$$

$$Q_1 = C_1 IA \qquad Q_2 = C_2 IA$$

$$96 = C_2 IA - C_1 IA$$

$$\frac{96}{0.35 * 160 - 0.15 * 160} = I = 3 \; in/hr$$

64.) If a paved concrete parking lot is designed, storm water best management practices could be used to do which of the following?

   a. Increase infiltration
   b. **Increase time of concentration**
   c. Increase peak discharge

d. Decrease rainfall intensity

Correct Answer (B)

Solution: Referencing the Fundamentals of Engineering Supplied-Reference Handbook; Civil Engineering Section; Environmental Engineering Section.

Storm water best management practices could be used to increase the time of concentration by creating a lesser slope in the parking lot. Infiltration will be minimal in a concrete parking lot. Increasing the peak discharge is not a best management practice. Decreasing the rainfall intensity has nothing to do with best management practices.

65.) What is the runoff in acre-feet for a 640 acre basin with a maximum basin retention of 0.5 inches for a 1 inch storm event?

a. 35 acre-feet
b. **30 acre-feet**
c. 25 acre-feet
d. 20 acre-feet

Correct Answer (B)

Solution: Referencing the Fundamentals of Engineering Supplied-Reference Handbook; Civil Engineering Section; Environmental Engineering Section.

The NRCS (SCS) Rainfall-Runoff equation is shown as;

$$Q = \frac{(P - 0.2S)^2}{P + 0.8S}$$

$$Q = \frac{(1 - 0.2 * 0.5)^2}{1 + 0.8 * 0.5} = 0.579 \; inches$$

Multiplying the runoff by the basin area, and converting to acre-ft gives us;

$$0.579 \; inches * \frac{ft}{12 \; inches} * 640 \; acre = \boldsymbol{30.9 \; acre - ft}$$

66.) A storm water conveyance channel should be used to deal with which type of flow?

   a. **Concentrated flow**
   b. Sheet flow
   c. Nutrient rich flow
   d. Base flow

   Correct Answer (A)

   Solution: Referencing the Fundamentals of Engineering Supplied-Reference Handbook; Civil Engineering Section; Environmental Engineering Section.

   Storm water conveyance channels should be used to deal with concentrated flow. Sheet flow should be dealt with using other best management practices. Nutrient rich flow does not make sense as an answer to this question. Base flow is not correct due to the fact that it occurs subsurface.

67.) The chart below shows the daily usage of water from a reservoir. What minimum volume should the reservoir be designed to store?

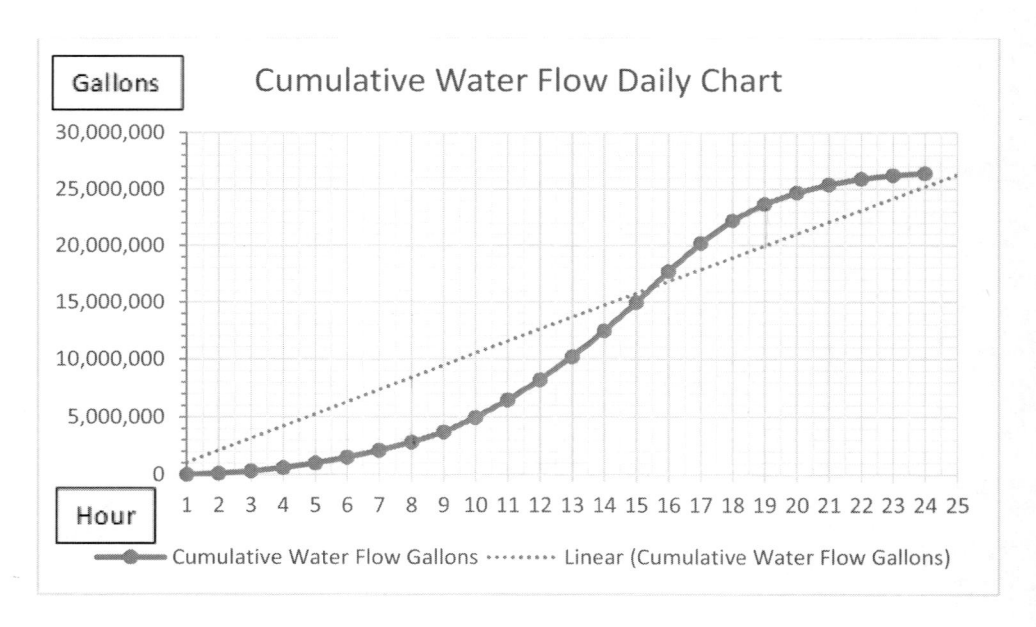

   a. 12 MG
   b. 26 MG
   c. 18 MG
   d. **9 MG**

Correct Answer (D)

Solution: Referencing the Fundamentals of Engineering Supplied-Reference Handbook; Civil Engineering Section; Environmental Engineering Section.

To solve this problem, you must find the maximum distance of the cumulative curve from the linear usage line at both the minimum and maximum.

$$= 5{,}000{,}000 \; at \; hour \; 8$$

$$= 3{,}500{,}000 \; at \; hour \; 19$$

$$\sum = 8.5 \; MG$$

68.) Excessive hardness of water in a potable water system can cause many issues. Potable water can be treated with which of the following to reduce hardness?

a. $Ca(OH)_2$
b. HCl
c. H_2SO_4
d. $(HCO_3)_2$

Correct Answer (A)

Solution: Referencing the Fundamentals of Engineering Supplied-Reference Handbook; Environmental Engineering Section.

In water softening, hydrated lime $[Ca(OH)_2]$ is used to precipitate the multivalent cations causing the hardness in the water. The hydrated lime is used in conjunction with soda ash $[Na_2CO_3]$.

69.) A raw water source flows into a circular tank where it is treated. The required hydraulic residence time is 1 hour, and the flow rate in is 100 gallons per minute. If the tank has a maximum height of 10 feet, what is the required diameter?

a. 8 feet
b. 5 feet
c. **10 feet**
d. 12 feet

Correct Answer (C)

Solution: Referencing the Fundamentals of Engineering Supplied-Reference Handbook; Environmental Engineering Section.

The equation for hydraulic residence time is shown as;

$$t_R = \frac{V}{Q}$$

Solving for Volume gives us;

$$(1\ hour)\left(100\ \frac{gallons}{minute}\right)\left(60\ \frac{minutes}{hour}\right)\left(\frac{ft^3}{7.48\ gallon}\right) = 802\ ft^3$$

$$V = \frac{\pi}{4} * d^2 * h$$

Solve for diameter in the equation;

$$\left[\frac{(802\ ft^3)(\frac{4}{\pi})}{10\ feet}\right]^{1/2} = 10.11\ feet$$

70.) A typical primary clarifier is needed to remove 64% of the suspended solids of the treated water. If the flow rate is 1 cfs, what is the necessary surface area of the clarifier?

a. 940 ft²
b. **810 ft²**
c. 1100 ft²
d. 850 ft²

Correct Answer (B)

Solution: Referencing the Fundamentals of Engineering Supplied-Reference Handbook; Environmental Engineering Section; Water Treatment Technologies.

From the Typical Primary Clarifier Efficiency Percent Removal Table, in order to remove 64% of the suspended solids the overflow rate must be 800 gpd/ft². The surface area can then be calculated using the following equation;

$$Overflow\ Rate = \frac{Q}{A_{SURFACE}}$$

Solving for surface area and plugging in the variables;

$$\frac{\frac{1 ft^3}{s} * 7.48 \frac{gallon}{ft^3} * \frac{86,400\ s}{day}}{800 \frac{\frac{gallon}{day}}{ft^2}} = 808\ ft^2$$

71.) A filtration system is used to treat water. The water has a terminal settling velocity of 13 m/s, and the porosity of the fluidized bed is 0.1. At what velocity should the system be backwashed?

a. 0.10 mm/s
b. 1.00 mm/s
c. **0.40 mm/s**
d. 0.60 mm/s

Correct Answer (C)

Solution: Referencing the Fundamentals of Engineering Supplied-Reference Handbook; Environmental Engineering Section; Filtration Equations.

The equation for the porosity of a fluidized bed is shown as;

$$\eta_{fb} = \left(\frac{V_B}{V_t}\right)^{0.22}$$

Solve for V_B by plugging in the known values.

$$(0.1)^{1/0.22} * 13\frac{m}{s} * 1000\frac{mm}{m} = 0.37\frac{mm}{s}$$

72.) A turbulent flow impeller is used in a rapid mix water treatment system. If the turbine has 6 curved blades and operates at a rotational speed of 100 rpm, how much power is needed to run the turbine? The diameter of the impeller is 0.2 m, the water density is 1000 kg/m³, and the flow is turbulent.

a. **7 W**
b. 8 W
c. 10 W
d. 68 W

Correct Answer (A)

Solution: Referencing the Fundamentals of Engineering Supplied-Reference Handbook; Environmental Engineering Section; Rapid Mix and Flocculator Design.

The Turbulent flow impeller mixer equation is shown as;

$$P = K_T(n)^3(D_i)^5\rho_f$$

From the table of Impeller Constants (K_T) the value for six curves blades is 4.80.

$$P = 4.80 * \left(100\ rpm * \frac{min}{60sec}\right)^3 (0.2\ m)^5 \left(1000\frac{kg}{m^3}\right) = 7.10\ W$$

73.) A potable water disinfection treatment is described by first order kinetics with a rate constant of 0.15 per minute. An ideal CMFR reactor is used, with a hydraulic residence time of 30 minutes. What is the effluent concentration if the initial concentration is 1000 mg/L?

a. **180 mg/L**
b. 1000 mg/L
c. 225 mg/L
d. 150 mg/L

Correct Answer (A)

Solution: Referencing the Fundamentals of Engineering Supplied-Reference Handbook; Environmental Engineering Section; Steady State Reactor Parameters Table.

The equation for a first order Ideal CMFR reactor is shown to be;

$$\theta = \frac{\left(\frac{C_0}{C_t}\right) - 1}{k}$$

Solving for C_t gives us;

$$C_t = \frac{C_0}{\theta k + 1} = \frac{1000 \ mg/L}{30 \ min * 0.15/\min + 1} = \mathbf{182 \ mg/L}$$

74.) A conventional activated sludge system treats 6 MGD. The aeration channel has a cross section of 30 ft by 11 ft. If the hydraulic residence time is 5 hours, what is the required length of the channel?

a. 600 ft
b. 700 ft
c. 650 ft
d. **500 ft**

Correct Answer (D)

Solution: Referencing the Fundamentals of Engineering Supplied-Reference Handbook; Environmental Engineering Section; Activated Sludge Section.

The equation for Hydraulic Residence Time is as follows;

$$\theta = \frac{V}{Q}$$

Solving for Volume gives us;

$$(5\ hour)\left(\frac{day}{24\ hour}\right)(6\ MGD)\left(10^6 \frac{gallon}{MG}\right)\left(\frac{ft^3}{7.48\ gallon}\right) = 167{,}112\ ft^3$$

Volume is equal to L*W*H; solve for length, since height and width are known.

$$\frac{167{,}112\ ft^3}{30\ ft * 11 ft} = 506\ ft$$

75.) An activated sludge aeration tank is 100 feet long, and 20 feet deep. The influent flow rate is 1.5 MGD, and the influent BOD is 160 mg/L. If the volumetric BOD_5 loading rate is 100 lb/(1000 ft^2*day), what is the necessary width of the tank?

a. 350 feet
b. **200 feet**
c. 275 feet
d. 180 feet

Correct Answer (B)

Solution: Referencing the Fundamentals of Engineering Supplied-Reference Handbook; Environmental Engineering Section; Water and Wastewater Treatment Section.

The first step is to find the mass inflow rate per day.

$$(1.5\ MGD)\left(160 \frac{mg}{L}\right)(8.34) = 2002\ lb/day$$

Dividing the mass by the loading rate gives us the volume of the tank.

$$\frac{(2002\frac{lb}{day})}{\left(\frac{100\ lb}{1000ft^2 * day}\right)} = 20,020\ ft^2$$

$$\frac{20,020\ ft^2}{100ft} = \mathbf{200\ ft}$$

76.) What is the solids loading rate in lb/day*ft² for activated sludge clarifiers with the following characteristics;

Number of units in parallel=2
Unit Diameter= 120 ft
Unit side water depth= 15 ft
Raw waste water inflow rate= 10 MGD
Return Activated Sludge flow rate= 3.0 MGD/clarifier
MLSS= 3500 mg/L
Raw Water BOD_5 = 225 mg/L
Raw Water SS = 265 mg/L

   a. **20**
   b. 18
   c. 16
   d. 7

Correct Answer (A)

Solution: Referencing the Fundamentals of Engineering Supplied-Reference Handbook; Environmental Engineering Section; Water and Wastewater Treatment Technologies; Activated Sludge.

The first step is to find the total flow rate. (Influent + return rate)

$$Q_T = 10\ MGD + 3\ MGD + 3\ MGD = 16\ MGD$$

Then determine the area of the two clarifiers.

$$A = 2 * \left[\frac{\pi}{4} * (120\ ft)^2\right] = 22{,}619\ ft^2$$

The solids loading rate can then be determined.

$$SLR = \left[\frac{16\ MGD * 3500\frac{mg}{L} * 8.34}{22{,}619\ ft^2}\right] = 20.65\frac{lb}{day * ft^2}$$

77.) A waste water treatment lagoon is designed to treat 1 MGD of wastewater with a concentration of 200 mg/L BOD₅ down to a concentration of 10 mg/L. The lagoon can be considered a second order ideal CMFR, with a rate coefficient of 3/day. What is the necessary volume of the lagoon?

a. 4,350,000 ft³
b. 107,000 ft³
c. **85,000 ft³**
d. 600,000 ft³

Correct Answer (C)

Solution: Referencing the Fundamentals of Engineering Supplied-Reference Handbook; Environmental Engineering Section; Steady State Reactor Parameters Section.

The equation for an ideal CMFR is shown as;

$$\theta = \frac{\frac{C_0}{C_t} - 1}{kC_t} = \frac{\frac{200\frac{mg}{L}}{10\frac{mg}{L}} - 1}{\frac{3}{day} * 10\frac{mg}{L}} = 0.63\ day$$

Using the equation for hydraulic residence time, the necessary volume can be calculated.

$$\theta = \frac{V}{Q} \qquad Q\theta = V$$

$$1\ MGD * 0.63\ day = 633{,}000\ gallon * \frac{ft^3}{7.48\ gallon} = 84{,}670\ ft^3$$

78.) An experimental air stripping water treatment system is designed with a liquid mole loading rate of 350 Kmol/s*m², and a height of 10 meters. What is the overall transfer rate coefficient?

a. 1.12/sec
b. 0.68/sec
c. 0.87/sec
d. **0.63/sec**

Correct Answer (D)

Solution: Referencing the Fundamentals of Engineering Supplied-Reference Handbook; Environmental Engineering Section; Water Treatment Technologies; Air Stripping.

The equation for height of a transfer unit is shown as;

$$HTU = \frac{L}{M_W K_L a} \qquad \frac{L}{HTU * M_W} = K_L a$$

$$\frac{350\ \frac{Kmol}{s * m^2}}{10\ m * 55.6\ \frac{Kmol}{m^3}} = \frac{0.63}{s}$$

79.) Determine the length of a rectangular clarifier if the flow rate is 5 MGD, the retention time is 2 hours, and the tank has a depth of 10 feet. The Length:Width ratio is 3:1.

   a. 112 feet
   b. **130 feet**
   c. 108 feet
   d. 100 feet

Correct Answer (B)

Solution: Referencing the Fundamentals of Engineering Supplied-Reference Handbook; Environmental Engineering Section; Water Treatment Technologies.

The equation for hydraulic residence time is;

$$t_R = \frac{V}{Q} \qquad Q * t_R = V = L * W * H$$

$$Qt_R = 5 \, MGD * \frac{10^6 \, gallon}{MG} * \frac{ft^3}{7.48 \, gallon} * \frac{day}{24 \, hour} * 2 \, hour = 55,704 \, ft^3$$

$$\text{since } 3W = L \qquad V = 3W * W * H$$

$$55,704 \, ft^3 = 3W * W * H$$

$$\frac{55,704 ft^3}{10 \, ft} = 3W^2 \qquad W = 43.09 \, ft \qquad 43.09 * 3 = L = \mathbf{130 \, ft}$$

80.) Determine the depth of the sorption zone on an activated carbon system with a total carbon depth of 10 feet, a volume of 2 MG at breakthrough, and a volume of 2.5 MG at exhaustion.

   a. 2.5 feet
   b. **2.2 feet**
   c. 3 feet
   d. 3.3 feet

Correct Answer (B)

Solution: Referencing the Fundamentals of Engineering Supplied-Reference Handbook; Environmental Engineering Section; Water Treatment Technologies; Activated Carbon Adsorption. The equation for depth of a sorption zone is as follow;

$$Z_s = Z \left(\frac{V_t - V_b}{V_T - 0.5 V_Z} \right)$$

Substituting the know variables into the equation gives us;

$$Z_s = 10 \, feet \left(\frac{2.5 \, MGD - 2.0 \, MGD}{2.5 \, MGD - 0.5(2.5 \, MGD - 2 \, MGD)} \right) = 2.2 \, feet$$

81.) Determine the minimum acreage needed to handle an inflow loading of 2750 lb BOD/day given the following design constraints; BOD max loading rate of 35 lb BOD/day/acre, and 0.2 lb BOD per 1000 ft³ with a max depth of 6 feet.

a. 70 acres
b. 34 acres
c. 53 acres
d. **79 acres**

Correct Answer (D)

Solution: Referencing the Fundamentals of Engineering Supplied-Reference Handbook; Environmental Engineering Section; Wastewater Treatment Technologies; Facultative Pond.

The first design constraint is 35 lb/day/acre, so the minimum number of acres can be calculated.

$$\frac{\left(\frac{2750\ lb\ BOD}{day}\right)}{\left(35\frac{lb}{day*acre}\right)} = 79\ acres$$

The acreage need for the second design constraint of 0.2 lb BOD per 1000 ft³ with a maximum depth of 6 feet can now be calculated.

$$\frac{\left(\frac{2750\ lb\ BOD}{day}\right)}{\left(0.2\frac{lb}{1000\ ft^3}\right)} = 13.75\ X\ 10^6\ ft^3$$

$$\frac{13.75\ X\ 10^6\ ft^3}{6\ feet\ max} * \frac{acre}{43,560\ feet^2} = 53\ acres$$

The larger value must be chosen to keep the facultative pond within all design constraints, so **79 acres** is the correct answer.

82.) An air sample shows a H₂S value of 75 ppm. The corresponding concentration in µg/L at 25°C and 1 atm of pressure is which of the following?

a. **104**
b. 78
c. 121
d. 10

Correct Answer (A)

Solution: Referencing the Fundamentals of Engineering Supplied-Reference Handbook; Thermodynamics Section; Ideal Gas Law.

Assuming 1 L of gas is present, the corresponding liters of H₂S can be calculated;

$$Volume\ H_2S = \frac{75\ ppm}{10^6} * 1.000\ L = 0.000075\ L$$

Using the Ideal Gas Law, the number of moles in 0.075 liter of gas can be determined.

$$PV = nRT \qquad n = \frac{PV}{RT}$$

$$n = \frac{(1\ atm)(0.000075\ L)}{(0.08206\frac{L*atm}{mol*K})(298\ K)} = 3.067 \times 10^{-6}\ mole$$

The concentration can then be determined by multiplying the number of moles by the molecular weight.

$$Conc = \frac{n*M_W}{V} = \frac{3.067 \times 10^{-6}\ mole * 34\frac{g}{mole} * \frac{10^6 \mu g}{g}}{1\ m^3} = 104\frac{\mu g}{m^3}$$

83.) Secondary air quality pollutants are primarily formed from the reactions of primary air quality pollutants. Where do these reactions generally occur?

  a. Combustion processes
  b. **Atmosphere**
  c. Pollution Control devices
  d. Production processes

Correct Answer (B)

Solution: Referencing the Fundamentals of Engineering Supplied-Reference Handbook; Environmental Engineering Section; Air Pollution Section.

Secondary air pollutants are generally formed by reactions of primary air quality pollutants that occur in the atmosphere. Primary air quality pollutants are generally emitted by production processes in a given facility.

84.) Coal that enters a power plant is pretreated to remove 98% of the mercury. If the coal has an average concentration of 1 g Hg/ton and the input rate for the power plant is 40,000 tons/hour, how much mercury is being emitted in g/s?

a. 0.11 g/s
b. 0.38 g/s
c. 800 g/s
d. **0.22 g/s**

Correct Answer (D)

Solution: Referencing the Fundamentals of Engineering Supplied-Reference Handbook; Environmental Engineering Section; Air Pollution Section.

The emission rate can be calculated per the following equation;

$$Emission = \left[40,000\frac{ton}{hr} * 1\frac{g\ Hg}{ton} * (1-0.98) * \frac{hr}{3600\ sec}\right] = \mathbf{0.22\ g/s}$$

85.) An incinerator has been tested and confirmed to remove 98% of all VOC's. The outlet concentration must be less than 50 µg/m³ and the maximum volumetric flow rate for the system is 10 m³/sec. What is the maximum flow rate of VOC's into the system in kg/day?

a. 1.3
b. **2.2**
c. 5.3
d. 6.8

Correct Answer (B)

Solution: Referencing the Fundamentals of Engineering Supplied-Reference Handbook; Environmental Engineering Section; Air Pollution Section.

The first step is to determine the mass flow rate out of the system at the maximum allowable concentration.

$$\dot{m}_{out} = Q * C_{OUT} = 10\frac{m^3}{s} * 50\frac{\mu g}{m^3} = 500\frac{\mu g}{s}$$

The equation for control efficiency is;

$$\eta = \frac{\dot{m}_{in} - \dot{m}_{out}}{\dot{m}_{in}} = 0.98 = \frac{\dot{m}_{in} - 500\frac{\mu g}{m^3}}{\dot{m}_{in}} = 0.98$$

$$\dot{m}_{in} = 25,000\frac{\mu g}{s}$$

$$25,000\frac{\mu g}{s} * \frac{kg}{10^9 \mu g} * \frac{86,400 sec}{day} = 2.16\frac{kg}{day}$$

86.) The stability class associated with the healthiest ambient air quality events is which of the following?

a. Stable
b. Slightly stable
c. Slightly unstable
d. **Very unstable**

Correct Answer (D)

Solution: Referencing the Fundamentals of Engineering Supplied-Reference Handbook; Environmental Engineering Section; Air Pollution Section.

Very unstable air conditions would promote the rapid rise of warmer air closer to the ground dispensing contaminates throughout the atmosphere and leading to lower concentrations. Lower concentration of contaminants would correspond to healthier air.

87.) A stack from an ore roasting operation emits CO at a rate of 20 g/s. The effective stack height is 100 m. What is the approximate maximum ground level concentration at a distance of 2 km downwind if the wind is blowing at 2.5 m/s on a sunny day with a few broken clouds?

a. 355 µg/m³
b. 2.5 µg/m³
c. 61 µg/m³
d. **34 µg/m³**

Correct Answer (D)

Solution: Referencing the Fundamentals of Engineering Supplied-Reference Handbook; Environmental Engineering Section; Air Pollution Section.

The equation for concentration downwind from an elevated source is shown as;

$$C_{(max)} = \frac{Q}{\pi u \sigma_Y \sigma_Z} exp\left[-\frac{1}{2}\frac{(H)^2}{\sigma_Z^2}\right]$$

The stability class can be found on the chart titled *Atmospheric Stability Under Various Conditions;* with a wind speed of 2.5 m/s on a sunny day with a few broken clouds the stability class is B.

From the Vertical Standard Deviations of a Plume Chart and the Horizontal Standard Deviations of a Plume chart, σ_Z and σ_Y can be found;

$$\sigma_Z = 225\ m \qquad \sigma_Y = 300\ m$$

$$C_{(max)} = \frac{20\frac{g}{s}}{\pi * 2.5\frac{m}{s} * 225\ m * 300\ m} exp\left[-\frac{1}{2}\frac{(100\ m)^2}{(225\ m)^2}\right] = 3.4178\ X\ 10^{-5}\frac{g}{m^3}$$

$$3.4178\ X\ 10^{-5}\frac{g}{m^3} * \frac{10^6 \mu g}{g} = 34.18\frac{\mu g}{m^3}$$

88.) Which pollutant would tend to travel the furthest in a plume?

a. **PM 2.5**
b. PM 10
c. PM 25
d. PM 100

Correct Answer (A)

Solution: Referencing the Fundamentals of Engineering Supplied-Reference Handbook; Environmental Engineering Section; Air Pollution Section.

In general the smaller the diameter of pollutant the larger the surface area to mass ratio. Thus the settling velocity is slowed more by frictional drag in a smaller molecule versus a larger molecule. With this in mind, the smallest diameter pollutant (PM 2.5), would tend to travel the furthest in a plume.

89.) A 40 foot tall scrubber treats a gas stream to a concentration of 100 µg/m³. If the height of a transfer unit is 2 feet and it has a stripping factor of 0.98, what is the inlet concentration of the gas stream?

a. 2000 µg/m³
b. 2250 µg/m³
c. **1750 µg/m³**
d. 20,000 µg/m³⁰

Correct Answer (C)

Solution: Referencing the Fundamentals of Engineering Supplied-Reference Handbook; Environmental Engineering Section; Air Pollution Section.

The first step is to calculate the number of transfer units (NTU).

$$Z = HTU(NTU) \qquad 40\ ft = 2ft * NTU \qquad NTU = 20$$

The inlet gas concentration then can be calculated by the following equation;

$$NTU = \left(\frac{R_S}{R_S - 1}\right) \ln \left(\frac{\left(\frac{C_{in}}{C_{out}}\right)(R_S - 1) + 1}{R_S}\right)$$

$$20 = \left(\frac{0.98}{0.98-1}\right) \ln\left(\frac{\left(\frac{C_{in}}{100\,\mu\frac{g}{m^3}}\right)(0.98-1)+1}{0.98}\right)$$

$$0.66 = \frac{\frac{C_{in}}{100} * -0.02 + 1}{0.98}$$

$$C_{in} = 1742 \frac{\mu g}{m^3}$$

90.) A high efficiency air cyclone has a body diameter of 0.5 m. It treats an air flow of 120 m³/min, at 350 K and 1 atm. What is the number of effective turns of the cyclone?

a. 4
b. 5
c. 6
d. 7

Correct Answer (C)

Solution: Referencing the Fundamentals of Engineering Supplied-Reference Handbook; Environmental Engineering Section; Air Pollution Section.

The equation to approximate a Cyclone's Effective Number of Turns is shown as;

$$N_e = \frac{1}{H}\left[L_b + \frac{L_c}{2}\right]$$

To calculate the variables in the equation, the Cycle Ratio of Dimensions to Body Diameter Table can be used for a high efficiency cyclone.

$H = 0.44 * 0.5\,m = 0.22\,m \quad L_B = 1.40 * 0.5\,m = 0.7\,m \quad L_C = 2.50 * 0.5\,m = 1.25m$

$$N_e = \frac{1}{0.22\,m}\left[0.7\,m + \frac{1.25\,m}{2}\right] = \mathbf{6.022}$$

91.) An Electrostatic Precipitator ("ESP") at a coal fired power plant must stay above 97% efficiency to stay in compliance. The ESP has 3 plates measuring 5 m by 10 m each, and the terminal drift velocity of the particles is 10 cm/s. What is the maximum allowable actual gas flow rate?

a. 2.6 m³/s
b. **8.6 m³/s**
c. 14 m³/s
d. 25.4 m³/s

Correct Answer (B)

Solution: Referencing the Fundamentals of Engineering Supplied-Reference Handbook; Environmental Engineering Section; Air Pollution Section.

The equation for ESP efficiency is as follows;

$$\eta = 1 - exp\left[-\frac{WA}{Q}\right] = 0.97$$

$$Q = -\frac{WA}{\ln(1-\eta)}$$

$$\frac{-0.10\frac{m}{s} * 5\ m * 10\ m * 3\ plates * 2\frac{sides}{plate}}{\ln(1-0.97)} = 8.56\frac{m^3}{s}$$

92.) Air flow in a baghouse at a cement plant has a temperature of 25°C and a flow rate of 10,000 scfm. Woven fabric bags are used to remove particulates from the air. Each bag is 9 inches in diameter and 10 feet long. How many bags are required to filter the air?

a. **216 bags**
b. 188 bags
c. 143 bags
d. 233 bags

Correct Answer (A)

Solution: Referencing the Fundamentals of Engineering Supplied-Reference Handbook; Environmental Engineering Section; Air Pollution Section.

Referencing the *Air-to-Cloth Ratio for Baghouses Table*, the flow rate per bag is equal to 0.6 m³/(min*m²). Converting this to English units gives us;

$$0.6 \frac{m^3}{m^2 * min} * 3.28 \frac{ft}{m} = 1.968 \frac{ft^3}{ft^2 * min}$$

The total bag area needed is then calculated as;

$$\frac{10,000 \; scfm}{1.968 \frac{ft^3}{ft^2 * min}} = 5081 \; ft^2$$

$$Area \; of \; a \; bag = \pi dh = \pi * \frac{9}{12} ft * 10 \; ft = 23.56 \frac{ft^2}{bag}$$

$$Number \; of \; bags = \frac{5081 \; ft^2}{23.56 \frac{ft^2}{bag}} = \textbf{216 bags}$$

93.) A wood treatment plant operates 360 days per year. The permit specifies that no more than 25 tons per year may be emitted of VOC's. If the emission factor is 5.8 X 10-3 lb/ft³ of treated wood, how many cubic feet can be treated in one year assuming no control devices are present?

a. 6.4 million
b. 7.8 million
c. 9.2 million
d. **8.6 million**

Correct Answer (D)

Solution: Referencing the Fundamentals of Engineering Supplied-Reference Handbook; Environmental Engineering Section; Air Pollution Section.

$$5.8 \; X \; 10^{-3} \frac{lb}{ft^3} * \frac{ton}{2000 \; lb} = 2.9 \; X \; 10^{-6} \frac{tons}{ft^3}$$

$$\frac{25\frac{tons}{year}}{2.9 \times 10^{-6}\frac{tons}{ft^3}} = 8,620,690\frac{ft^3}{year}$$

94.) If an environmental assessment (EA) is performed for a proposed project and it is determined to have the possibility to cause significant impact, what is the next step in the permitting process?

    a. **Environmental Impact Study**
    b. Remediation
    c. Delineation
    d. Baseline sampling

Correct Answer (A)

Solution: Referencing the Fundamentals of Engineering Supplied-Reference Handbook; Environmental Engineering Section.

An environmental assessment is performed to determine if a project could possibly cause significant impact to the environment. If it is deemed possible that the project could cause significant impact, an Environmental Impact Study must be performed to further determine the proposed project's impact.

95.) A community generates 75,000 lb/day of solid waste that is deposited in a municipal landfill. The permit for the landfill stipulates a ratio of refuse to cover of 1:2. The in place density of the material (both refuse and cover) is 1200 lb/yd³. If the operational permit is for 10 years, how much material (yd³) does the landfill need to be designed to hold?

    a. **684,000**
    b. 732,000
    c. 61,000
    d. 183,000

Correct Answer (A)

Solution: Referencing the Fundamentals of Engineering Supplied-Reference Handbook; Environmental Engineering Section.

The landfill is assumed to be operating 365 days per year.

$$\frac{Refuse\ weight + Cover\ weight}{\gamma_{Material}} = Volume$$

$$\frac{\left[\frac{75,000\ lb}{day} * \frac{365\ day}{year} * 10\ year\right] + \left[\frac{75,000\ lb}{day} * \frac{365\ day}{year} * 10\ year * \frac{2}{1}\right]}{1200\ lb/yd^3} = \mathbf{684,375\ yd^3}$$

96.) A landfill is limited to a depth of 60 feet. The landfill serves a community of 110,000 people, with a waste generation rate of 4 lb/day per person. If the density of the compacted waste is 1000 lb/yd³ and the cover material account for 20% of the total volume, what amount of acreage must be disturbed? Assume a 20 year landfill life.

   a. 3.3 acre
   b. 49 acre
   c. **42 acre**
   d. 6.5 acre

Correct Answer (C)

Solution: Referencing the Fundamentals of Engineering Supplied-Reference Handbook; Environmental Engineering Section.

The first step is to determine the total amount of waste that will be generated;

$$\frac{\frac{4\ lb}{person * day} * 110,000\ people * \frac{365\ day}{year} * 20\ year}{1000\ \frac{lb}{yd^3}} = 3,210,000\ yd^3$$

Accounting for the cover material the total volume is;

$$\frac{3,210,000\ yd^3}{0.80} = 4,015,000\ yd^3$$

Dividing the volume by the maximum allowable depth then converting to acres gives us;

$$A = \frac{V}{d} = \frac{4,015,000 yd^3}{60ft * \frac{yd}{3ft}} * \left(\frac{3 ft}{yd}\right)^2 * \frac{acre}{43,560 ft^2} = 41.5\ acres$$

97.) Which of the following class of compounds could be safely combined with hydrazine?

   a. Carbamates
   b. Esters
   c. **Caustics**
   d. Cyanides

   Correct Answer (C)

   Solution: Referencing the Fundamentals of Engineering Supplied-Reference Handbook; Environmental Engineering Section; Hazardous Waste Compatibility Chart.

   Carbamates + Hydrazine= Heat and gas generation (BAD)

   Esters + Hydrazine = Heat and gas generation (BAD)

   Caustics + Hydrazine = No adverse consequences (GOOD) **Correct Answer**

   Cyanides + Hydrazine = Gas generation (BAD)

98.) An industrial facility has the following hazardous wastes on site; water reactive substances, aromatic hydrocarbons, amides, and cyanides. What is the minimum number of containers needed to store the waste?

   a. 1
   b. **2**
   c. 3
   d. 4

   Correct Answer (B)

   Solution: Referencing the Fundamentals of Engineering Supplied-Reference Handbook; Environmental Engineering Section; Hazardous Waste Compatibility Chart.

Water Reactive substances are extremely reactive, and should not be mixed with any other chemicals (1 container needed).

Aromatic hydrocarbons can be mixed with all substances generated on this site (2 containers needed). Amides can be mixed with cyanide, and that can be mixed with aromatic hydrocarbons. Two total containers will be necessary to safely store the wastes.

99.) A clay liner is being installed in a landfill and will be compacted to a porosity of 0.20, a hydraulic conductivity of 10^{-5} ft per day, and a thickness of 5 feet. If the hydraulic head is 15 feet, what is the breakthrough time in years?

   a. 88
   b. 94
   c. **69**
   d. 54

Correct Answer (C)

Solution: Referencing the Fundamentals of Engineering Supplied-Reference Handbook; Environmental Engineering Section; Landfill Section.

The equation for Break-Through Time for Leachate to Penetrate a Clay Liner is shown as;

$$t = \frac{d^2 \eta}{K(d+h)} = \frac{(5\ ft)^2 * 0.20}{\left(\frac{10^{-5} ft}{day}\right)\left(\frac{365 day}{year}\right)(5\ feet + 15\ feet)} = 68.5\ years$$

100.) Which of the following is a characteristic that is used to classify waste as hazardous?

   a. Oxidative
   b. Inertivity
   c. **Reactivity**
   d. Leachability

Correct Answer (C)

Solution: Referencing the Fundamentals of Engineering Supplied-Reference Handbook; Environmental Engineering Section; Hazardous Waste.

Reactivity, or the potential for violent reactions, is one of the four categorizing characteristics for hazardous waste along with corrosivity, ignitability, and toxicity.

101.) After initial compaction, waste at a landfill has a density of 800 lb/yd3. After the waste is placed in the landfill, cover material is placed on top until the overburden pressure on the waste is 200 psi. What is the specific weight of the material at this pressure? Assume empirical constants of 0.1 yd³/in², and 5 X 10⁻⁴ yd³/lb.

a. **1800 lb/yd³**
b. 1600 lb/yd³
c. 1900 lb/yd³
d. 1300 lb/yd³

Correct Answer (A)

Solution: Referencing the Fundamentals of Engineering Supplied-Reference Handbook; Environmental Engineering Section; Landfill Section.

The equation for the Effect of Overburden Pressure is shown as;

$$SW_p = SW_i + \frac{p}{a+bp} = 800 \; lb/yd^3 + \frac{200 \frac{lb}{in^2}}{0.1 \frac{yd^3}{in^2} + 5 \times 10^{-4} \frac{yd^3}{lb} * 200 \frac{lb}{in^2}} = \mathbf{1800 \; lb/yd^3}$$

102.) A cover layer on a landfill has undergone a change in the amount of water that it holds in storage. The change is measured by saturation testing as a decrease of 2.0 inches per unit area. If the layer has undergone precipitation of 3 inches, runoff of 1.2 inches, and evapotranspiration of 0.3 inches, how much water per unit area has percolated into the waste below?

a. 1.5 in
b. 2.0 in
c. 2.5 in
d. **3.5 in**

Correct Answer (D)

Solution: Referencing the Fundamentals of Engineering Supplied-Reference Handbook; Environmental Engineering Section; Landfill Section.

The Soil Landfill Cover Water Balance equation is shown as;

$$\Delta S_{LC} = P - R - ET - PER_{sw}$$

Since the problem asks for the amount of water that has percolated through the landfill cover, we can solve the equation for PER$_{sw}$ and plug in the known variables.

$$PER_{sw} = P - R - ET - \Delta S_{LC} = 3 - 1.2 - 0.3 - (-2) = \mathbf{3.5\ inches}$$

103.) The activity of a radionuclide was measured after 30 days at 600 Bq. If the half-life is known to be 11 days, what was the initial activity level?

a. 100 Bq
b. **4000 Bq**
c. 2000 Bq
d. 3000 Bq

Correct Answer (B)

Solution: Referencing the Fundamentals of Engineering Supplied-Reference Handbook; Environmental Engineering Section; Radiation Section.

The equation for half-life is shown to be;

$$N = N_0 e^{-\frac{0.693t}{\tau}}$$

Solving for N₀ gives us;

$$N_0 = \frac{N}{e^{-\frac{0.693t}{\tau}}} = \frac{600 Bq}{e^{(\frac{-0.693*30}{11})}} = 4000\ Bq$$

104.) A saturated clay loam soil has a porosity of 35%. If the specific gravity of the clay is 2.80, what is the specific weight of the saturated soil?

a. 163 lb/ft³
b. **135 lb/ft³**
c. 120 lb/ft³
d. 118 lb/ft³

Correct Answer (B)

Solution: Referencing the Fundamentals of Engineering Supplied-Reference Handbook; Environmental Engineering Section.

$$Porosity = \frac{V_{VOID}}{V_{TOTAL}}; assume\ 1\ ft^3\ is\ the\ volume$$

$$V_{VOID} = (0.35)(1\ ft^3) = 0.35\ ft^3 \qquad V_{SOLID} = 1 - 0.35\ ft^3 = 0.65\ ft^3$$

$$Weight_{Total} = (0.35 ft^3)(\gamma_{Water}) + (0.65 ft^3)(SG_{SAND})(\gamma_{Water})$$

$$W_T = 0.35 * 62.4 \frac{lb}{ft^3} + 0.65 * 2.8 * 62.4 \frac{lb}{ft^3} = 135.4 \frac{lb}{ft^3}$$

105.) Using the Dupuit formula, what flow rate from a well (cfs) is required to draw down an aquifer 20 feet from the base of the aquifer a ¼ mile away? It is determined that the well draws down the aquifer 8 feet from the base an 1/8 mile away at the same flow rate. Assume K is equal to 10⁻³ ft/sec.

a. **1.5 cfs**

b. 18 cfs
c. 12 cfs
d. 0.09 cfs

Correct Answer (A)

Solution: Referencing the Fundamentals of Engineering Supplied-Reference Handbook; Civil Engineering Section; Environmental Engineering Section; Well Drawdown.

Dupuit's formula is shown as;

$$Q = \frac{\pi k(h_2^2 - h_1^2)}{\ln\left(\frac{r_2}{r_1}\right)}$$

$$Q = \frac{\pi * \frac{10^{-3} ft}{sec} * (20^2 - 8^2)}{\ln\left(\frac{1320\ ft}{660\ ft}\right)} = 1.5\ cfs$$

106.) If a partially confined aquifer has a hydraulic conductivity of 3x10⁻² cm/s, a thickness of 35 m, and a storage coefficient of 0.20, what is the transmissivity of the aquifer?

a. 585 m²/day
b. 1115 m²/day
c. 680 m²/day
d. **910 m²/day**

Correct Answer (D)

Solution: Referencing the Fundamentals of Engineering Supplied-Reference Handbook; Civil Engineering Section; Environmental Engineering Section.

$$Transmissivity = Hydraulic\ Conductivity * Aquifer\ thickness$$

$$T = 3\ X\ 10^{-2} \frac{cm}{s} * \frac{m}{100cm} * \frac{86,400 sec}{day} * 35\ m = \mathbf{907} \frac{m^2}{day}$$

107.) A confined aquifer is 30 m thick and has a width of 0.5 km. Two observation wells are located 0.8 km apart in the direction of flow. The head in well #1 is 50 m, and in well #2 it is 38 m. Hydraulic conductivity is 0.5 m/day in the aquifer and the effective porosity is 0.3. What is the flow rate of water in the aquifer?

a. **115 m³/day**
b. 95 m³/day
c. 85 m³/day
d. 105 m³/day

Correct Answer (A)

Solution: Referencing the Fundamentals of Engineering Supplied-Reference Handbook; Civil Engineering Section; Environmental Engineering Section.

$$Q = -KA \frac{h_2 - h_1}{L} \qquad A = 30\ m * 500\ m = 15{,}000\ m^2$$

$$Q = -\frac{0.5\ m}{d} * 15{,}000\ m^2 * \left[\frac{38\ m - 50\ m}{800\ m}\right] = \frac{113\ m^2}{day}$$

108.) An aquifer drops 2 feet in elevation over a distance of 1350 feet. A tracer is deployed in the aquifer, and takes 8 months to travel 300 feet. What is the hydraulic conductivity of the aquifer in feet per second?

a. 10^{-1}
b. 10^{-2}
c. **10^{-3}**
d. 10^{-4}

Correct Answer (C)

Solution: Referencing the Fundamentals of Engineering Supplied-Reference Handbook; Civil Engineering Section; Environmental Engineering Section.

Darcy's Law is shown to be;

$$Q = -KA\left(\frac{dh}{dx}\right)$$

Solving for K gives us;

$$K = -\frac{Q}{A * \frac{dh}{dx}} \quad since \; \frac{Q}{A} = V \; the \; equation \; changes \; to;$$

$$K = \frac{-V * dx}{dh} = \frac{\left(\frac{300 ft}{8 \, month}\right)\left(\frac{month}{30 day}\right)\left(\frac{day}{86400 \, s}\right) 1350 \, ft}{2 \, feet} = 9.8 \times 10^{-3} ft/sec$$

109.) An aquifer has a thickness of 50 feet, over an area of 10 acres. If the porosity of the aquifer is 0.4 and the saturation is 90%, how much water is contained in the aquifer?

a. 170 acre-ft
b. **180 acre-ft**
c. 190 acre-ft
d. 225 acre-ft

Correct Answer (B)

Solution: Referencing the Fundamentals of Engineering Supplied-Reference Handbook; Civil Engineering Section; Environmental Engineering Section.

$$Aquifer \; Volume = A * B * \eta * sat = 10 \; acre * 50 \; feet * 0.4 * 0.9 = \mathbf{180 \; acre - ft}$$

110.) In which part of the groundwater system are the pore spaces not filled with water?

a. Capillary Fringe
b. **Zone of Aeration**
c. Zone of saturation
d. Hydraulic gradient

Correct Answer (B)

Solution: Referencing the Fundamentals of Engineering Supplied-Reference Handbook; Civil Engineering Section; Environmental Engineering Section.

The zone of aeration is the unsaturated area above the aquifer in which the pore spaces are not filled with water. The Capillary Fringe is the saturated zone just above the water table. The zone of saturation has its pore spaces filled with water. While the hydraulic gradient is the slope of the water table.

Made in United States
North Haven, CT
02 February 2022